100.00

369 0289966

D1758604

ATLAS OF DEVELOPMENTAL FIELD ANOMALIES OF THE HUMAN SKELETON

ATLAS OF DEVELOPMENTAL FIELD ANOMALIES OF THE HUMAN SKELETON

A PALEOPATHOLOGY PERSPECTIVE

Ethne Barnes

WILEY-BLACKWELL

A John Wiley & Sons, Inc., Publication

Published by John Wiley & Sons, Inc., Hoboken, New Jersey.
Published simultaneously in Canada.

Library of Congress Cataloging-in-Publication Data:

Barnes, Ethne.
  Atlas of developmental field anomalies of the human skeleton : a paleopathology perspective / by
Ethne Barnes.
      p. cm.
  Includes index.
  ISBN 978-1-118-01388-5 (cloth)
  1. Paleopathology–Atlases. 2. Skeleton–Abnormalities–Atlases. I. Title.
  R134.8B36 2013
  611'.70222–dc23

                                                                                          2012016832

Printed in the United States of America.

10 9 8 7 6 5 4 3 2 1

Dedicated to Charles F. Merbs

CONTENTS

C. RIBS

D. STERNUM

PREFACE

While studying nonmetrical traits in the human skeleton, I began to question the how and why of their development. This led to a study of human skeletal embryology that revealed the various developmental fields responsible for the construction of the human skeleton. Thus, the morphogenetic approach to the analysis of developmental anomalies began with my Arizona State University dissertation, published in 1994, "Developmental Defects of the Axial Skeleton in Paleopathology," by the University Press of Colorado.

Over the years, I have tested this approach to the analyses of developmental anomalies over and over again with positive results, adding just a few revisions and additions for the axial skeleton. However, I have seen the need for clarification of this method into a more simplified version from the earlier text, plus the need to include the appendicular skeleton. Too often there has been questioning, confusion between diseased or traumatized bones and developmental disorders, and misunderstanding of how the various skeletal anomalies develop, and whether or not the morphogenetic approach is applicable. Recent genetic studies within molecular biological embryology provide the necessary proof of the genetic components governing the expressions of skeletal anomalies, supporting the morphogenetic methodology. Molecular DNA studies have revealed the complex interaction of genetic signaling along specified genetic pathways and the genetic variations leading to developmental anomalies within specific developmental fields in the embryo. Altered genetic signaling can affect how a skeletal part is structured, including programming for secondary ossifications.

Thus, the genetic interactions within developmental fields of the evolving embryo set the stage for anomalous or defective development, including skeletal structural variations. Each developmental field is governed by its own set of genetic instructions that can be altered by mutant genetic signals or epigenetic interference at specific developmental threshold events. The outcome is deviation from the expected construct. More than one disturbance can occur within the same developmental field as different threshold events take place. Sometimes an upset in one developing field impinges upon another developmental field, leaving both with anomalous results. Most developmental disturbances follow familial genetic linkages. Some developmental field disturbances appear with specific groups of disorders known as syndromes.

The medical community has labored for decades with developing and refining classifications for specific congenital skeletal defects based on observations of autopsy specimens and radiographs, sometimes with confusing results. Paleopathologists have the advantage of observing dry bone specimens from thousands of human skeletons, thus providing a different perspective to the study of developmental anomalies within the skeleton that appear along a gradient of variations for each anomaly not usually seen in the medical community. This allows paleopathologists to "think outside the box" of medical constructs with a different approach to understanding and classifying skeletal anomalies.

Development of skeletal anomalies cannot be too simplified as there is much variation within variations. Thus, orderly classification according to developmental fields allows for defining the many variations occurring along the same theme, including anomalies yet to be identified. I am constantly amazed at the variable expressions of many anomalies that come to my attention. Even when there is more than one type of anomaly occurring in the same developmental field or in adjacent fields, it is possible to sort them out accordingly. For example, it is not unusual to find more than one type of developmental disturbance in the same vertebral column but all can be sorted out by the morphogenetic approach. And since no two multiple vertebral disturbances are alike, this can be very informative.

While costly ancient DNA studies have recently commanded the attention of research in population studies, attention has been diverted from empirical genetic analyses of past human skeletal populations. The morphogenetic approach to genetic studies in skeletal populations, while not replacing ancient DNA studies, remains a useful methodology particularly where invasive studies are not permitted. Patterns of data collected by the morphogenetic approach alone can help identify human migrations, genetic drift, marriage patterns, and familial linkages.

The research for the morphogenetic approach to the skeletal analyses of developmental anomalies is built on the works of many researchers, some going back over 100 years. Although it is not possible to cite them all, I am grateful for their contributions that have helped shape this approach. The questioning of how and why developmental anomalies occur began with my studies of nonmetric traits under Michael Finnegan at Kansas State University. I would not have discovered the world of vertebral anomalies without the guidance of Charles F. Merbs at Arizona State University. Thus, the concept of the morphogenetic approach was born.

Dave Hunt and Don Ortner at the National Museum of Natural History (NMNH) also played a vital role in this process. Troy Case at North Carolina State University provided much needed information on the developmental anomalies of hands and feet. I also want to thank Kristen Parlstein and Kathleen Adia at NMNH for their kind help locating specimens for this book. Most of all, I may not have had the courage to pursue this endeavor if not for the constant encouragement and support of my loving husband, Art Rohn.

Ethne Barnes

LIST OF FIGURES

INTRODUCTION

The purpose of this text is to provide an easy reference guide for the identification and understanding of developmental field variations of skeletal structure, both pathological and nonpathological forms in paleopathology. Anomalies in this text are defined as structural bone variants deviating from designated standard ranges, excluding metabolic defects in bone tissue known as skeletal dysplasias (see Aufderheide and Rodriguez-Martin 1998; Ortner 2003). This text primarily focuses on those anomalies most likely to be found in human skeletal remains. However, variations within variations of each developmental field skeletal anomaly do occur, and hopefully, enough basic information is presented as a guide for identifying the categories of structural anomalies for such variants and to understand how they develop.

Evolutionary principles govern variation within gene pools. Evolution cannot exist without genetic variation, thus the potential for a wide range of developmental variability of skeletal anomalies exists in all populations. Most of these variables do not threaten overall function and are easily maintained within a gene pool. Since variability can be expressed in many forms, it is the type of expression in variation that reflects underlying genetics. Thus, it is the underlying genetic base that determines different types of variable mutant genetic expressions. Some populations may harbor mutant genes allowing for very rare anomalies not seen in other populations, while some more common variants can be found in several populations but may be absent or far less frequent in a single population. The overall frequency pattern of anomalies provides a non-invasive view of the genetic pattern of a population, and can reveal familial linkages and marriage patterns and help track migrations. Similarities in the overall pattern of expressed anomalies between populations suggest a shared gene pool, while differences reflect genetic drift. Patterns of expressed variations are best represented by percentages based on the number of specific expressions against the number of surviving bony elements (not the number of individuals) where the variation may occur. Very rare anomalies can be significant even at very low frequencies versus significantly high frequencies for commonly occurring variants.

Variation in skeletal structures reflect disturbances during morphogenesis within the underlying embryonic developmental field, usually resulting from genetic mutations (variations) acting on a susceptible genetic base during a critical threshold event (Barnes 1994a; Larsen 2001; Sadler 2006). Occasionally, other factors such as maternal infection, exposure to detrimental environmental contaminants/drugs, or nutritional disorders can behave as epigenetic factors acting like mutant genes on critical threshold events in embryonic development. Rapid change marks a critical threshold event, usually when newly formed cells are proliferating, migrating, or differentiating. Mutant genes or epigenetic influences can only act on susceptible genetic backgrounds during critical threshold events during morphogenesis, and most embryos with severe developmental disturbances rarely survive the first trimester

Atlas of Developmental Field Anomalies of the Human Skeleton: A Paleopathology Perspective, First Edition. Ethne Barnes.
© 2012 Wiley-Blackwell. Published 2012 by John Wiley & Sons, Inc.

of pregnancy. Morphogenesis marks the essence of organ development and sets the pattern for continued bone growth into maturity.

The morphogenetic approach to analysis of skeletal anomalies addresses variability that occurs within the various developmental fields that produce different parts of the skeleton during embryonic development (Barnes 1994a, 2008, 2012). Analysis of variations in skeletal development utilizing this approach is very effective in defining and categorizing the range of developmental anomalies, from minor to major expressions, occurring throughout the skeleton. It remains as an open-ended approach that can explain any kind of variation of a localized developmental disturbance in the human skeleton. Many minor cranial variations have previously been described as summarized by Hauser and De Stefano (1989).

Unlike the bone tissue-specific disturbances or dysplasias (Ortner 2003) such as the chondrodysplasias, this type of analysis focuses specifically on the developmental fields within the evolving embryo that produce the various structural precursors of the human skeleton (Gruneberg 1963, 1964). A developmental field is defined as the close interaction of select developing embryonic tissues involved in the complex composition of a specific structure or set of closely related structures. Sometimes a number of disturbances linked by their timing to critical threshold events can affect more than one developmental field to present as a combined set of defects known medically as a syndrome. Sometimes disturbance in one developing field can interfere with development in an adjacent developmental field.

The embryonic membranous skeletal precursors forming in the first weeks of life determine the fate of finished bony elements. The pattern for development is set and cannot be altered. Recent molecular genetic studies have revealed that the driving forces of morphogenesis are controlled by a complex of regulatory genes producing various transcription proteins that activate cascading genetic signals for various complex sequential events along developmental pathways, resulting in the construction of the various primordial tissues as they come together. Disturbances within the molecular regulatory genetic signaling pathways have been identified in association with variations in organ development, including the varied parts of the skeleton (Larsen 2001; Sadler 2006).

The timing via molecular signaling within and between specific developing tissues coming together signify threshold events that are crucial for normal sequential development as embryonic tissues organize into specific developmental fields responsible for the construction of particular organs or structures. Most developmental variants result from delayed timing of threshold events via genetic mutation altering molecular

signaling pathways, or influences from similar disturbances in adjacent developmental fields.

Much of the human skeleton derives from paired primordial tissues that develop concurrently, such as the limbs, the two sides of the face and skull, both sides of the rib cage, vertebrae, and sternum. Delay in timing on one side of a structure or paired set of structures will result in asymmetrical growth. Asymmetrical faces and limbs are not uncommon. Primordial segmentation also occurs in developing limbs, cranial and vertebral parts, sternum, and ribs. Delay, disorder, or absence of segmentation creates abnormal development.

The human skeletal structure is divided into the axial and appendicular skeletons. The 80 bones of the axial skeleton include the facial, cranial, and ear bones of the skull; and the hyoid, vertebral column, ribs, and sternum of the thorax. The appendicular skeleton represents the 126 bones of the upper limbs—shoulder girdles, arms, wrists and hands, and lower limbs—pelvis, legs, and feet, plus very small sesamoid ossicles developing within the tendons of the hands and feet. The primordial blastema of the skeletal structures initially develops from the embryonic membranous tissue derived from mesenchymal cells originating in the mesoderm germ layer responsible for the various connective tissues. Bones of the face, skull cap, clavicles, and digit tips of the primordial skeleton ossify directly from membranous tissue, while the rest of the skeleton transforms into cartilaginous tissue before ossifying.

The precursors of the axial skeleton appear first at the beginning of the third week following the formation of the primitive streak in the midline of the germ disk (Fig. IA). The primitive streak marks the boundary for the bilateral symmetrical development of the two sides of the body with the future head at the slightly elevated top of the streak, the primitive node, surrounding a depression known as the primitive pit (Fig. IB). Epiblast cells migrating to the primitive streak receive signals from the streak to proliferate and detach from other epiblast cells and migrate through the streak to differentiate into the germ layers endoderm and mesoderm, while the remaining epiblast cells form the third germ layer, the ectoderm.

Some mesenchymal cells migrate to the head end to form the prechordal plate responsible for the induction of the forebrain, followed by the formation of a dense midline tube just caudal to the prechordal plate. As this tube grows in length while the primitive streak regresses, it transforms into a solid rod—the notochord. The notochord acts as the scaffolding necessary for the development of the axial skeleton. Mesenchymal cells migrating from the head end line up on both sides of the notochord to form a matching pair of cylindrical condensations known as paraxial mesoderm—precursors of the vertebral column with contributions to the base of the skull. With the completion of the vertebral structures, the notochord will regress but leave behind rudimentary tissue that gives rise to the formation of the nucleus pulposus in the center of the vertebral disks.

The paired columns of paraxial mesoderm form the forerunners of the vertebral column, undergoing sequences of cranial–caudal metamorphosis beginning with transformation into paired whorl-like segments—somitomeres, with the more cranial ones contributing to the formation of the head. Those developing sequentially from the occipital region of the head and down along each side of the notochord segment into pairs of tissue blocks called somites (Figs. I and II): 4 occipital, 8 cervical, 12 thoracic, 5 lumbar, 5 sacral, and 8–10 coccygeal pairs. The first occipital and last five to seven coccygeal segments are absorbed into adjacent somite tissue.

Each somite block separates into sclerotome, myotome, and dermatome tissue, with the ventromedial sclerotomes developing into primordial vertebrae. Sclerotome cells migrate from the paired somites to surround the notochord, meeting and uniting midline to form the tissue block precursors of the vertebral bodies, while dorsal sclerotome cells surround the developing neural tube and merge to form the precursor vertebral neural arch complex. Cells for each vertebral part respond to signals from adjacent tissue structures in their development. Segmentation plays a major part in the development of the vertebral column beginning with the somites, and resegmentation as the sclerotomes split into cranial and caudal halves, with the caudal halves fusing with the cranial halves of succeeding sclerotomes. Resegmentation allows for the development of the fibrous intervertebral disks that separate the developing vertebral segments.

The ribs of the thorax develop from the primordial thoracic costal processes, while the mesosternum evolves from paired mesenchymal tissue condensation bands or bars that form on the ventrolateral sides of the body wall, moving ahead of the developing ribs. As the uppermost part of the sternal bands meet midline, they unite with a small midline condensation of mesenchymal cells—the precostal process, joined by a pair of small mesenchymal condensations—suprasternal structures—that form the interface between the developing clavicles and manubrium. The sternal bands continue to merge midline in a cranial–caudal direction, with the caudal ends the last to meet. Once the sternal bands unite to form the primordial sternum, it segments into the manubrium and four mesosternal sections—sternebrae, with the remaining caudal tissue remnants forming the xiphoid process.

Bilateral thickening of the epiblast induced by the underlying prechordal plate and cranial portion of the notochord creates the neural plate (Fig. IC,D)—precursor to the central nervous system. The cranial end broadens out into paired broad folds to create the two sides of the brain, while the narrower caudal ends gradually fold toward each other and fuse into the developing neural tube of the future spinal cord underlying the notochord. Fusion of the neural tube begins at the cranial end, gradually moving caudally with the extension of the precursor structures of the axial skeleton. The open cranial end—cranial neuropore, remains open until a stage of the upper portion of the developing axial skeleton is reached, followed in a few days by closure of the open caudal end—caudal neuropore in the primordial lumbar–sacral region, sealing the tubular structure of the future spinal cord with the developing brain (Fig. IE,F). This is followed with progressive caudal canalization within the caudal eminence of the tail end of the embryo to complete the lower end of the primordial neural tube.

Cells along the lateral borders of the broad neural folds of the cranial end of the neural plate undergo transformation into neural crest cells and leave the neural folds. Some of these cells migrating before neural tube closure contribute to the blastemal desmocranium, precursor of the cranium, and the pharyngeal (branchial) arches (Figs. IF and IIA–C) that form the craniofacial skeleton and structures within the neck. The base of the skull arises separately from a cartilaginous plate forming from the nasal area to the anterior end of the neural tube.

As development of the axial skeleton progresses, the appendicular skeleton begins with the appearance of paired tissue buds forming from lateral plate mesoderm condensations (Fig. II). The upper limb buds appear first in the cervical–thoracic region under one set of master genes followed by the lower limb buds in the lumbar–sacral region programmed from another set of master genes. Each limb bud consists of an outer ectodermal cap and inner mesodermal core. Apical ectodermal tissue of the growing limb bud forms a thickened ridge—the apical ectodermal ridge (AER) on the tip end of the bud with bordering mesenchymal tissue that remains undifferentiated—the progress zone (PZ). The AER and PZ work together for the outgrowth of the limb along the proximal–distal axis. Genetic signals from a cluster of cells on the posterior border of the limb bud near the AER contribute to positioning and pattern formation of the limbs along the anterior–posterior axis, especially the digits, while genetic signals from areas of overlying ectoderm maintain polarity of limb formation along the dorsoventral axis. As the limb grows, cells furthest from the AER begin a sequence of developmental events leading to the different segments of the limbs, while the terminal ends flatten out into disk-shaped plates—the hand and foot plates. Fingers and toes arise from the ridges within

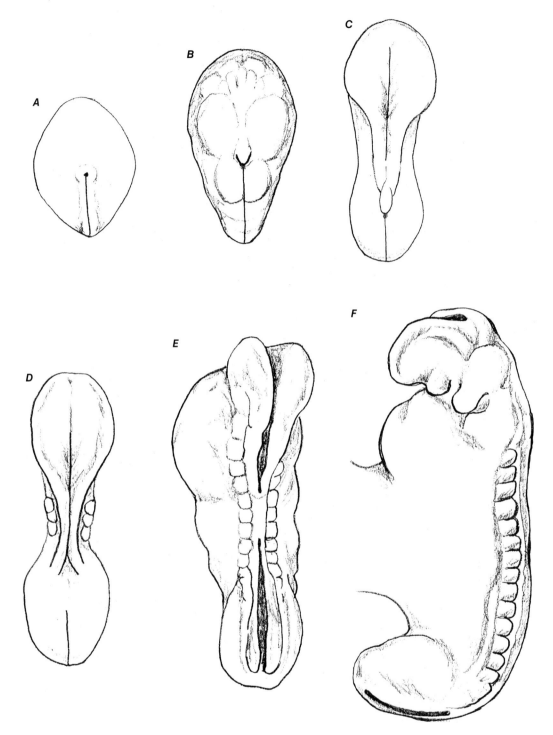

FIGURE I. Embryonic development: (*A*) 16 days and (*B*) 18 days—primitive streak and node; (*C*) 19 days—neural (prechordal) plate elevating to form neural folds around the neural groove; (*D*) 20 days—neural tube forming with paired somites developing within the paraxial mesoderm; (*E*) 22 days—continued development of somites within the paraxial mesoderm along the sides of lengthening neural tube as neural crests form; (*F*) 25 days—cranial and caudal ends of the neural tube (neuropores) remain open, bulges for pharyngeal arch I (maxilla and mandible) with the first pharyngeal groove (external auditory meatus) and pharyngeal arch II (stylohyoid chain) appear, caudal eminence pronounced.

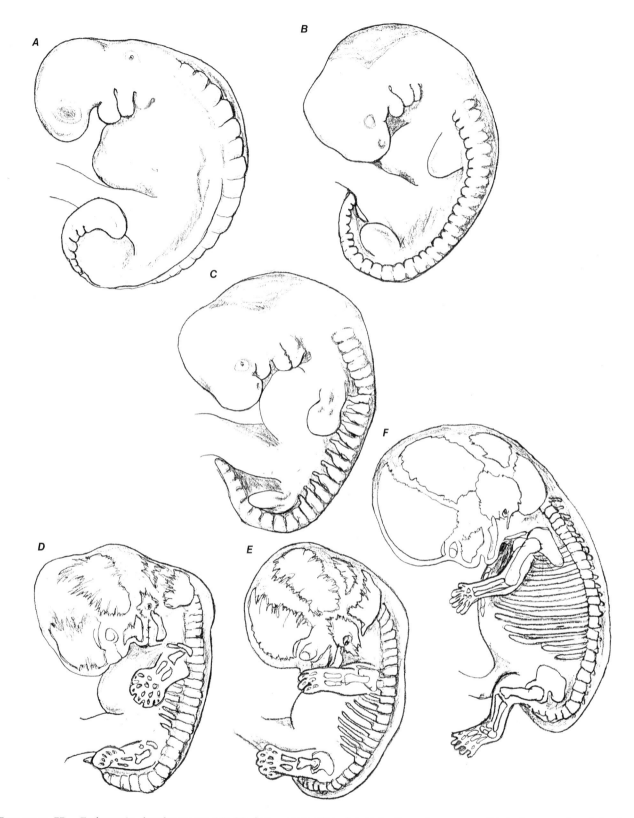

FIGURE II. Embryonic development: (*A*) 30 days—optic placodes and otic nodes appear, third pharyngeal arch begins to appear, thickening of the ridge for the upper limb bud showing, caudal tail pronounced; (*B*) 34 days—nasal placodes and limb buds develop; (*C*) 37 days—hand plates, rib development begins; (*D*) 42 days; (*E*) 46 days; (*F*) 60 days—membranous skeleton completed.

the plates that separate into distinct rays of tissue by programmed cell death of intervening tissue. Limb formation consists of continuous mesenchymal condensations that segment into the membranous bone models that in turn develop into hyaline cartilaginous bone precursors, separated by arrested primordial cartilage condensations designated as interzones for the precursor joints.

By the end of 8 weeks, the embryonic membranous skeleton is complete, providing the template for the developing bony skeleton and ossification via membranous or cartilage models begins. Completion of the membranous skeleton also includes programmed timing for the development of primary and secondary ossification centers and the final fusion of epiphyses in the mature skeleton.

The following pages present an easy reference for identifying and defining recognized structural skeletal anomalies from head to toe, beginning with the axial skeleton and ending with the appendicular skeleton. Each part of the skeleton is outlined and discussed according to embryonic developmental fields involved, accompanied by definitions and illustrations of skeletal anomalies within their specific developmental fields. Each section follows a particular part of the skeleton— A through F—with associated numbers for each topic covered in that section, and the numbering system matches the figure numbers for the associated illustrations. Also, the NMNH reference in captions stands for the National Museum of Natural History at the Smithsonian.

Since expressions of any developmental field anomaly can have considerable variation, and more than one type of developmental upset can affect a single field, the morphogenetic approach remains open ended. This allows for the interpretation of unusual anomalies not addressed or as yet unrecognized using this atlas as a guide.

AXIAL SKELETON

CHAPTER A

SKULL

A-1. CRANIAL VAULT DEVELOPMENT

The skull vault arises from the embryonic blastemal (membranous) desmocranium surrounding the developing brain. Neural crest cells are induced by overlying ectodermal tissue to move from the edges of each side of the neural plate folds to form mesenchymal condensations adjacent to specific areas of the developing brain, beginning with the forebrain. Each mesenchymal condensation is associated with a specific part of the brain. By the thirteenth day, these mesenchymal condensations transform into membranous curved plates on each side of the growing brain to form the primordial cranial vault: frontal, parietals, and interparietal squamosa of the occipital bone (Fig. A-1.0). By the seventh week, these bones begin to ossify directly as bony spicules that radiate outward from developing osseous centers within the membranous tissue. The frontal bone ossifies from a pair of osseous centers, one on each side separated by the fetal metopic suture that normally fuses as the two halves grow together after birth by age 2. Each parietal generally ossifies from two osseous centers, one above the other. The squamosal or interparietal portion of the occipital ossifies from a complex of ossification centers with the number varying according to genetic instructions. As the interparietal occipital ossifies, it fuses at the mendosa line with the expanding lower portion of the occipital of the chondocranium as it ossifies from the cartilage.

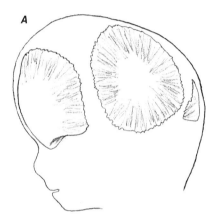

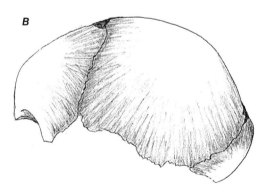

FIGURE A-1.0. Calvaria development: frontal, parietals, occipital interparietal or squamosa—(*A*) embryonic membranous bones; (*B*) newborn bones.

Atlas of Developmental Field Anomalies of the Human Skeleton: A Paleopathology Perspective, First Edition. Ethne Barnes.
© 2012 Wiley-Blackwell. Published 2012 by John Wiley & Sons, Inc.

The primitive cranial bones are separated by seams and spaces of connective tissue—sutures and fontanelles. These separations allow the cranial bones to be molded during descent through the birth canal. Postnatal cranial bone growth closes the small fetal sagittal fontanelle before birth. The posterolateral (mastoid) fontanelles close by the end of the first year. The posterior and anterolateral (sphenoid) fontanelles close by 3 months and the anterior fontanelle by 18 months.

CRANIAL VAULT ANOMALIES

A-1.1. Extra Ossicles

These commonly form within sutures, especially the lambdoidal suture. Less often they develop within the fontanelles (Fig. A-1.1). Multiple and variable extra ossicles (especially along the lambdoidal suture) also occur with cleidocranial dysostosis, a disturbance in the

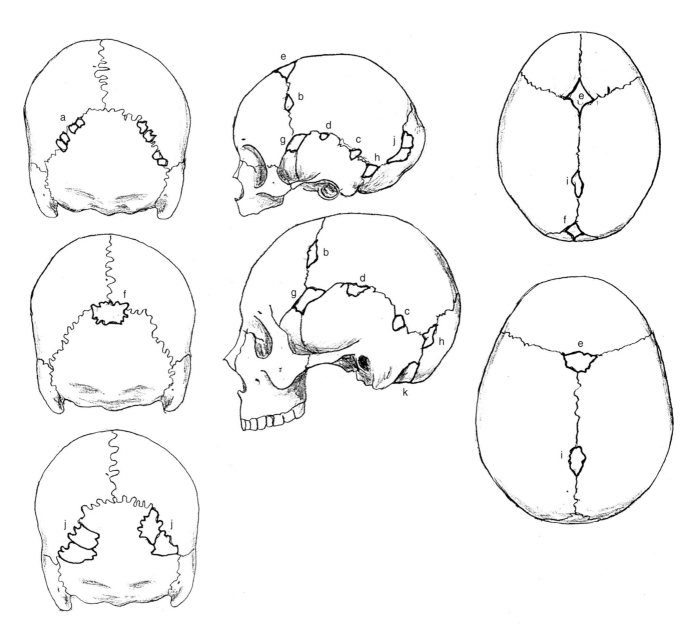

FIGURE A-1.1. Extra ossicles: (newborn and adult) (a) lambdoidal ossicles; (b) coronal ossicles; (c) parietal notch ossicle; (d) temporal squamosa ossicle; (e) bregma (anterior fontanelle) ossicle; (f) lambda (posterior fontanelle) ossicle; (g) epipteric (anteriolateral fontanelle) ossicle; (h) asterion (posterolateral fontanelle) ossicle; (i) obelion (fetal sagittal fontanelle) ossicle; (j) occipital interparietal ossicle; (k) retromastoid ossicle within the chondocranium between the mastoid and occipital.

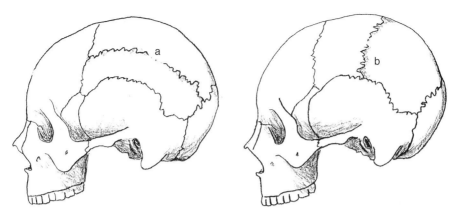

FIGURE A-1.2.1. Extra parietal sutures: (a) horizontal; (b) vertical (Anderson 1995; Shapiro 1972).

membranous bone tissue development that also affects the membranous development of the clavicle.

A-1.2. Extra Sutures

These usually result from the failure of the membranous parts of the same primordial cranial bone to coalesce completely or partially prior to ossification, isolating separate ossification centers within the bone (Fig. A-1.2.1). The metopic suture dividing the infant frontal usually grows together but often remains in place throughout life. The fetal mendosa line between the membranous squamosa and cartilaginous occipital base that normally disappears before birth sometimes remains in place as a complete or incomplete suture. The persistent complete suture gives the appearance of an extra large interparietal occipital bone, commonly referred to as the inca bone (Fig. A-1.2.2).

A-1.3. Sutural Agenesis

This is the failure of sutures to develop (completely or partially) between opposing membranous cranial bones. The lack of bony separation can lead to various forms of cranial deformation, especially with more than one type of suture affected (Figs. A-1.3.1–A-1.3.3).

A-1.4. Parietal Thinning

This is the failure of diploe space to develop within the superior–posterior region of the parietal. This creates a somewhat ovoid depression on the surface of the cranium (Fig. A-1.4.1). It has been noticed in children as well as adults. Most often it appears bilateral, but can be seen unilateral, and the affected area is quite thin, becoming more pronounced with age (Fig. A-1.4.2).

A-1.5. Enlarged Parietal Foramina

This is the failure of membranous bone development within the region of the parietal foramina outlets for Santorini's emissary veins, often associated with some form of lambdoidal and or sagittal suture agenesis (Hoffman 1976). The defects are covered with fibrous membrane instead of bone, ranging in shape from large ovoid to elongated slit bony openings (Figs. A-1.5.1–A-1.5.3).

A-1.6. Inclusion Cysts

These form during embryogenesis with failure of overlying primordial ectodermal cells to retreat from developing cranial bones, usually occurring midline anywhere between the frontonasal region and foramen magnum of the skull base. The anterior fontanelle and the area near the occipital protuberance are the most common sites, sometimes forming in the roof of the eye orbit or greater wing of the sphenoid (Fig. A-1.6.1). Size varies from a few millimeters up to around 10 cm. Tiny cysts often go undetected. Depending on the timing of the event of epidermal entrapment, developing cysts can derive from undifferentiated epidermal cells alone or with differentiated deeper dermal cells. Epidermoid cysts generally contain keratohylin within a capsule of stratified squamosal epithelium and continue to grow slowly by desquamation. Dermoid cysts may contain oil, sebum, cholesterol, and hair follicles, and increase slowly in size by glandular secretions along with epithelial desquamation. Both types can occur internally or externally, leaving cystic depressions within the affected bone. Epidermal cysts frequently occur within the dura mater covering of the brain, while dermoid cysts usually develop within the cranial diploe or between the periosteum and scalp. The cystic bony imprint

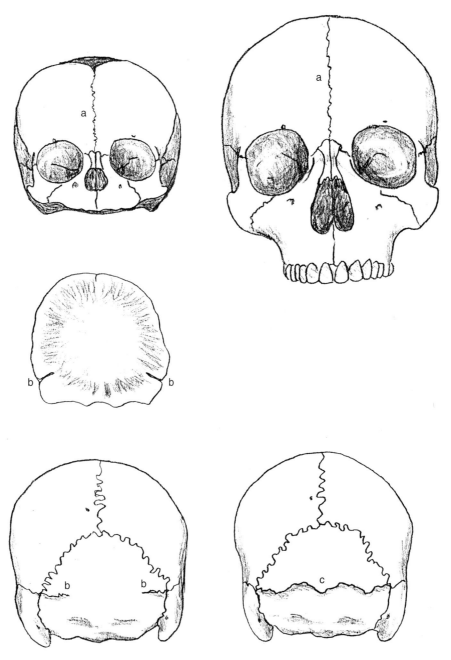

FIGURE A-1.2.2. Extra sutures: (newborn and adult) (a) metopism—retention of infantile suture; (b) remnant fetal mendosa suture; (c) complete retention of fetal mendosa suture (also known as inca bone).

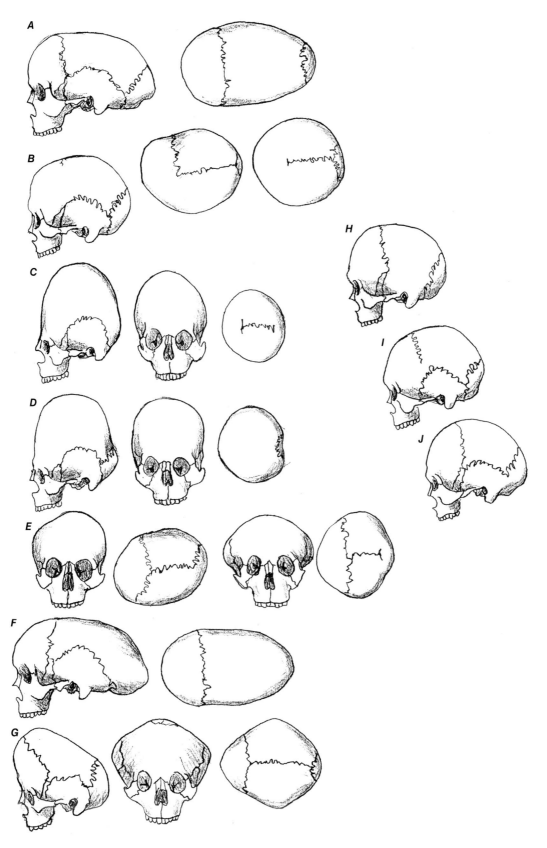

FIGURE A-1.3.1. Sutural agenesis: (*A*) sagittal suture agenesis—scaphocephalic deformity; (*B*) coronal suture agenesis—unilateral or partial, may have plagiocephalic deformity, bilateral—brachycephalic form; (*C*) coronal and lambdoidal suture agenesis—oxycephalic (tower skull) deformity; (*D*) coronal and sagittal suture agenesis—oxycephalic (tower skull) deformity with bulging occipital; (*E*) lambdoidal suture agenesis—unilateral or partial, may have plagiocephalic deformity, bilateral—brachycephalic form; (*F*) lambdoidal and sagittal suture agenesis—dolichocephalic form; (*G*) metopic suture agenesis—trigonocephalic deformity; (*H*) temporal squamosa suture agenesis—no deformity; (*I*) sphenofrontal suture agenesis—no deformity; (*J*) temporoccipital suture agenesis—no deformity.

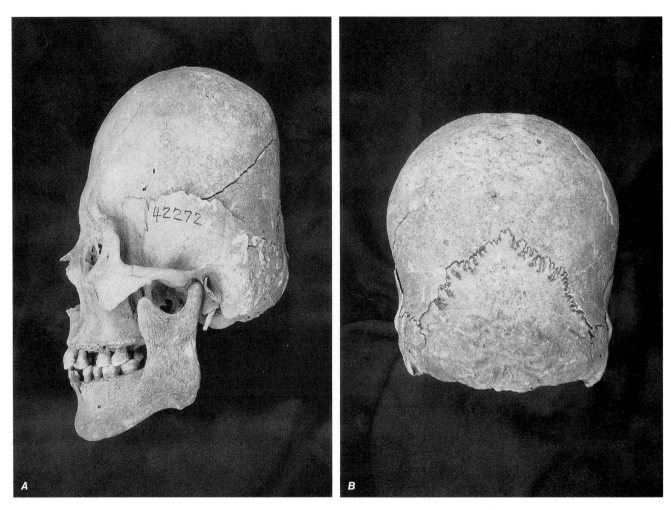

FIGURE A-1.3.2. Sutural agenesis—oxycephaly: coronal and sagittal sutures absent, adult female, Little Colorado River, AZ; (*A*) lateral and (*B*) occipital views (Field Museum).

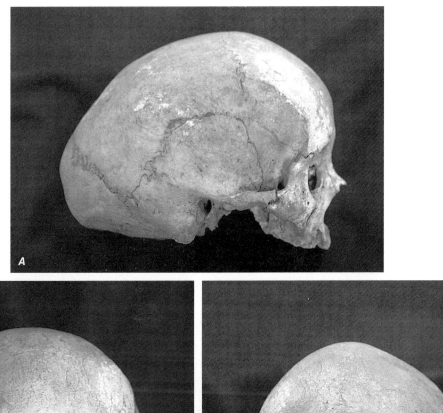

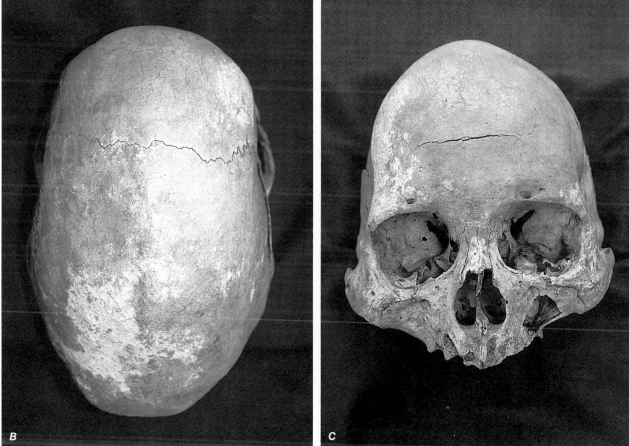

FIGURE A-1.3.3. Sutural agenesis—scaphocephaly: sagittal suture absent, adult male (NMNH 293841), Cerros, Peru; (*A*) lateral, (*B*) top, and (*C*) facial views.

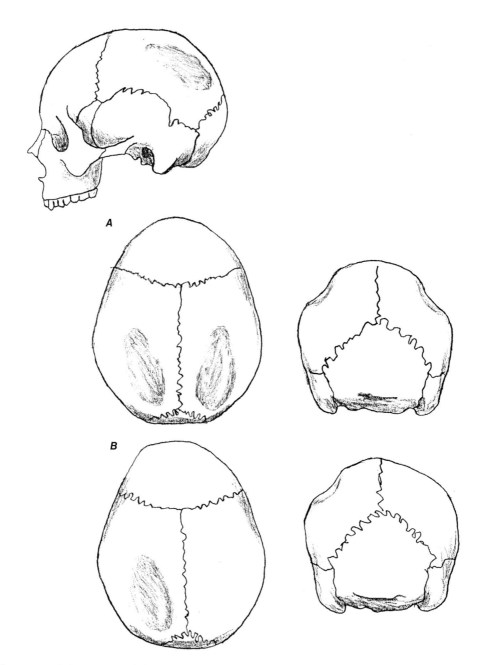

F I G U R E A-1.4.1. Parietal thinning: (*A*) bilateral; (*B*) unilateral.

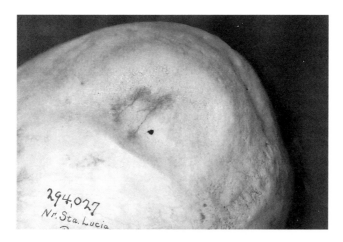

FIGURE A-1.4.2. Parietal thinning close-up: right side, adult male (NMNH 294027), N. Sta. Lucia, Peru.

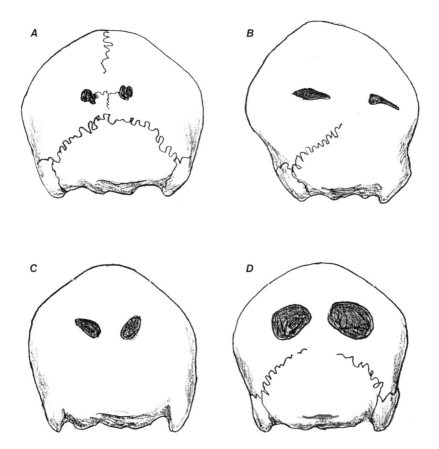

FIGURE A-1.5.1. Enlarged parietal foramina: (*A*) cruciform shape with localized sagittal suture agenesis; (*B*) slit shape with coronal, sagittal, and unilateral right lambdoidal suture agenesis; (*C*) lozenge shape with sagittal and lambdoidal suture agenesis; (*D*) large ovoid shape with partial sagittal and lambdoidal suture agenesis (Hoffman 1976).

FIGURE A-1.5.2. Enlarged parietal foramina: adult male (NMNH 276981) with agenesis sagittal and lambdoidal sutures, Ponce Mound, Santa Clara county, CA.

FIGURE A-1.5.3. Enlarged parietal foramina slits: 4-year-old child (NMNH 276982) with agenesis sagittal suture, right side of the lambdoidal suture, and midportion (bregma area) of the coronal suture, near Palo Alto, CA.

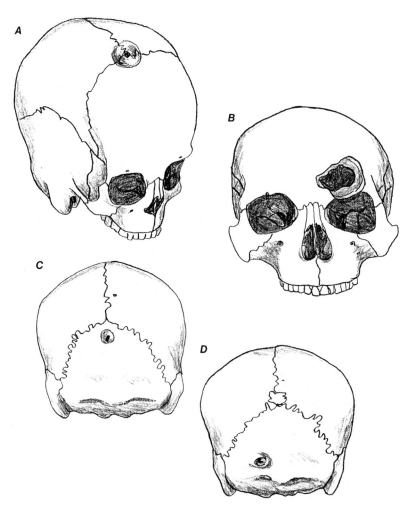

FIGURE A-1.6.1. Cranial inclusion (dermoid) cysts: (*A*) at the bregma; (*B*) involving the supramedial margin left eye orbit; (*C*) below the lambda; (*D*) above the nuchal ridge, left of center.

leaves a rounded depression with a sharp border, often with a thin bony floor (Fig. A-1.6.2). Sometimes an external dermoid cyst communicates with an internal ectodermal cyst by way of a dermoid sinus through a rounded opening in the floor of the external cyst, penetrating through all layers of the bone (Fig. A-1.6.3), or ending blindly within the bony diploe (Rubin et al. 1989; Scheie and Albert 1977).

A-1.7. Cranial Neural Tube Defects

These occur with neurulation disturbances in the cranial end of the developing neural folds and tube. If the neural folds fail to fuse, the developing brain and upper spinal cord develop abnormally with exposure to the amniotic fluid—craniorachischisis, and the stimulus for the corresponding bone development is impaired, including cervical and thoracic neural arch development that leaves them widely spaced in a flayed appearance.

Fetal death usually occurs with this disorder in the first trimester (Dudor 2010).

Developmental failure of the anterior neuropore to form causes failure of the brain tissue to form above the brain stem—anencephaly, resulting in death before or shortly after birth. Development of the membranous cranial vault stimulated by the developing brain is severely impaired with severe reduction and deformity of the bones within the desmocranium. The developing chondocranial base is also affected with deformities, especially midline structures (Dudor 2010).

Developing brain tissue can also be disrupted with delay in closure of the anterior neuropore, a postneurulation defect. This allows brain tissue and or the meningeal brain covering to protrude through an opening of overlying developing bone tissue. The abnormally placed tissues are encased in the covering epidermal tissue, appearing as a skin-covered cyst. Whether brain tissue or just the meningeal brain

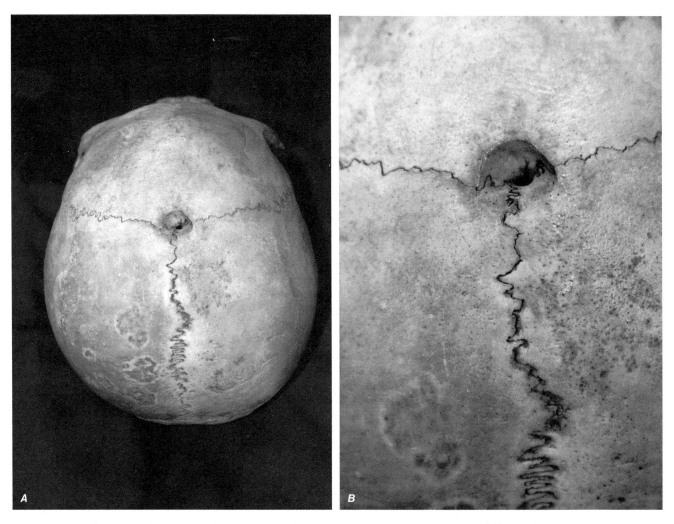

FIGURE A-1.6.2. Cranial inclusion (dermoid) cyst at the bregma: (*A*) adult female (NMNH 264629), Chicama, Peru; (*B*) close-up (Barnes 1994:53; I mistakenly identified this a meningocele).

covering is involved in the defect depends on the timing of the delay. When only the meninges protrude through the adjacent developing bone, it is known as a meningocele, and when the brain tissue is involved, it is an encephalocele. They usually develop along the sagittal plane from the nasion root to the base of the occipital, sometimes occurring in the roof of the orbital angle or root of the sella turcica (Fig. A-1.7.1). Individuals with a meningocele can survive into adulthood, while those born with an encephalocele usually do not survive infancy. Encephaloceles generally develop at the base of the skull and can be quite large with the affected bone appearing bifurcated (Lemire 1988). The anterior fontanelle or bregma region is a common site for a meningocele, leaving a depressed cystic impression of varying size on affected bone with an irregular opening through the bony floor (Fig. A-1.7.2). The saucer-shaped or rounded bony depression has well-defined raised borders surrounded by an outer flange of bony buildup

responding to pulsations from the meningeal tissue (Webb and Thorne 1985).

A-1.8. Hydrocephaly

This is an abnormally enlarged cranium caused by increased accumulation of cerebrospinal fluid within the ventricles of the brain resulting from interference with its normal circulation and absorption throughout the spinal and cerebral chambers. This can be due to developmental disturbances, particularly spinal neural tube defects (spina bifida) or defects in the subarachnoid spaces or aqueduct of Sylvius. Disturbances can also arise before or after birth from injury, infection, or brain tumors. Therefore, developmental cause and effect may be difficult to establish. Developmentally related hydrocephalics can survive beyond birth. Normal neonate cranial circumference ca. 35 cm will increase almost 1 cm per month until reaching ca. 46 cm at 1 year while

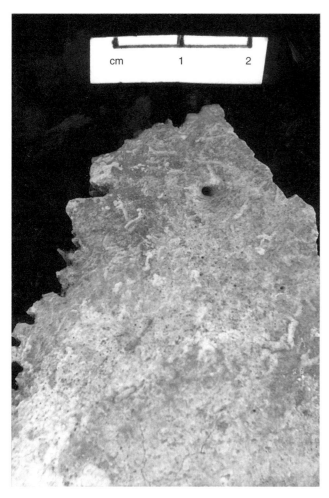

FIGURE A-1.6.3. Cranial inclusion (dermoid) cyst near the lambda: occipital of an adult male, Frankish Corinth, Greece.

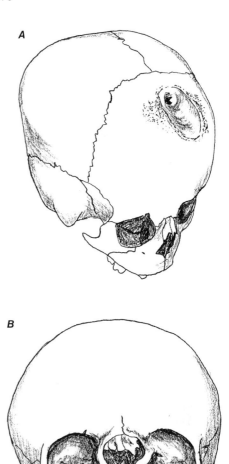

FIGURE A-1.7.1. Cranial neural tube defects: (*A*) frontal meningocele; (*B*) nasal region encephalocele (drawn from Ortner 2003:455).

the hydrocephalic skull will be much larger (Warkany 1971). The cranium is unusually large and globular in shape with enlarged anterior fontanelle that fails to close. The face appears disproportionately small with roof of eye orbits appearing raised upward (Fig. A-1.8). Signs of developmental defects affecting the aqueduct of Sylvius with inadequate circulation of fluid can be delayed until early adolescence.

A-1.9. Microcephaly

An abnormally small cranium resulting from a defective developing small brain is known as microencephaly, with the forebrain and occipital lobes primarily affected. The cranial sutures usually remain in place as with normally developing calvaria. Head circumference in affected older children and adults falls below 46 cm (about the size of a 1-year-old) with brain weight less than 900 g (Goodman and Gorlin 1983; Warkany 1971). The cranium has a conoidal ("pinhead") shape from narrow, receding frontal and shortened occipital that is

often flattened (Fig. A-1.9.1). The face appears large compared with the abnormally small-sized skull (Fig. A-1.9.2). The affected individual suffers severe mental retardation (idiot) and can live into adulthood, able to perform simple duties or tricks. Microcephaly can be familial.

A-2. FACE DEVELOPMENT

The membranous bones of the upper midface evolve from the bulging viscerocranium below and forward of the evolving blastemal desmocranium, as neural crest cells migrate into the region from the mid- and hind-brain areas. The center swells to form the frontonasal prominence responsible for the upper midregion of the

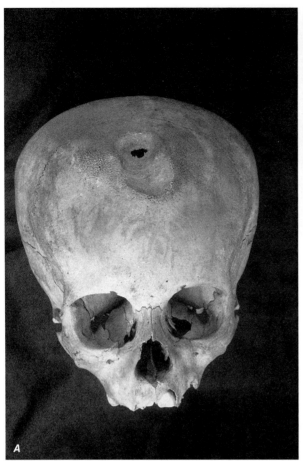

FIGURE A-1.7.2. Meningocele neural tube defect: (*A*) near the bregma on frontal, 7- to 8-year-old child, Ancon, Peru (Field Museum); (*B*) close-up.

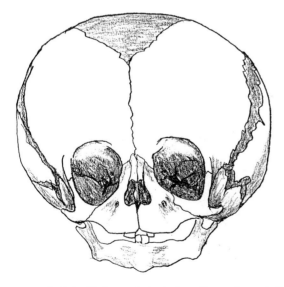

FIGURE A-1.8. Hydrocephaly: infant (drawn from Aufderheide and Rodriguez-Martin 1998:57).

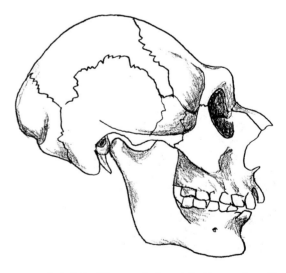

FIGURE A-1.9.1. Microcephaly: adult (drawn from Brothwell 1981:169).

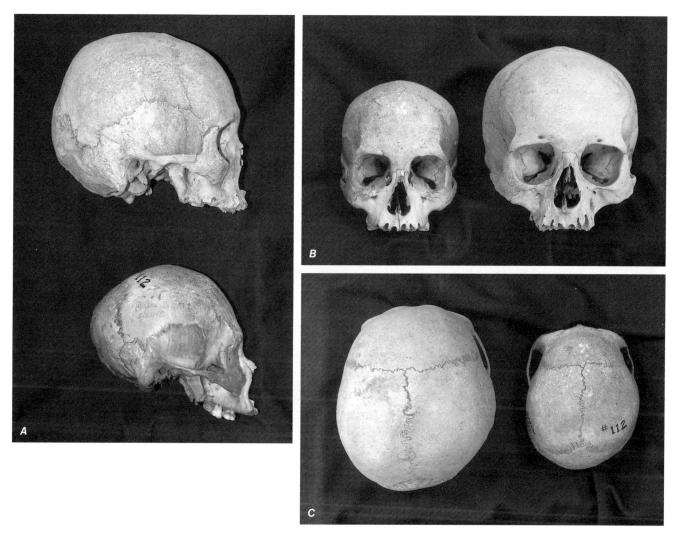

FIGURE A-1.9.2. Microcephaly: adolescent female (NMNH 379510), Chicama, Peru, with normal adult female skull. (*A*) lateral, (*B*) faces, and (*C*) skull tops.

face, while the first pair of pharyngeal or branchial arches appearing along each side (Fig. II) forms the lateral and lower portions of the face—the maxilla, mandible, zygomatics as well as the temporal squamosa, parts of the sphenoid, palatine bones, and malleus and incus ear bones. By the end of the fourth week, the face begins to take shape around the primitive mouth (stomodeum) as the five merging developmental fields—the frontonasal prominence and paired maxillary and mandibular prominences from the first pharyngeal arch—come together. Paired nasal placodes soon appear on the frontonasal prominence to form the nasal pits with raised rims dividing the prominence into paired lateral and medial nasal parts. The medial nasal parts immediately join with the formation of the primordial nasal bones and bony septum formed by the ethmoid and vomar bones within the primitive nasal cavity. The inferior portion expands bilaterally below

the nasal cavity to form the intermaxillary process, where the paired halves of the bony premaxilla take shape and fuse together. The premaxilla unites with the two halves of the primordial maxilla to form the upper jaw and palate. Ectodermal grooves develop between the lateral nasal processes and maxillary prominences as they move toward each other, forming nasolacrimal grooves that become the nasolacrimal ducts bounded by the lacrimal bones as the membranous bones meet and fuse together (Fig. A-2.0.1). Between the eighth and ninth weeks the medial walls of the maxillae extend downward to form palatine shelves as the developing tongue moves downward, joining with the small primary palatal extension from the premaxilla. As the two halves expand and come together, they form the primordial membranous bony palate (Fig. A-2.0.2) while continuing to position into place and join with the horizontal plates of the palatine bones. The posterior portion of

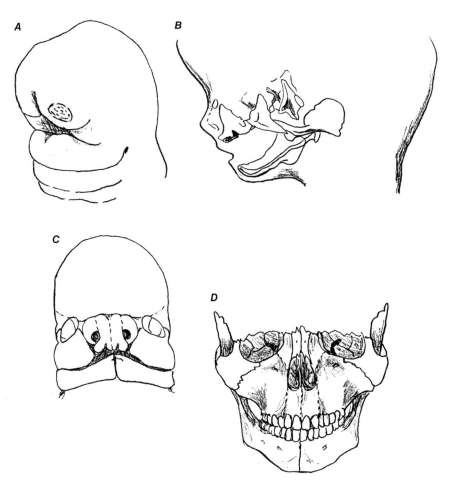

FIGURE A-2.0.1. Facial development: (*A*) frontonasal prominence with the left nasal placode, paired maxillary and mandibular prominences from the first pharyngeal arch surround the primitive mouth (30 days); (*B*) corresponding developing membranous facial bones (9 weeks); (*C*) frontonasal prominence subdivides into medial and lateral nasal parts as maxillae move toward the center and mandibular parts come together (7 weeks); (*D*) corresponding developmental fields (dotted lines) within a mature bony face.

the palatine bones does not ossify but develops into the soft palate and uvula.

Development of the two halves of the mandible depends on the placement of cartilaginous bars known as Meckel's cartilage, extending throughout each mandibular prominence, attracting neural crest cells to form the membranous mandibular bony tissue around them (Fig. A-2.0.1B). Six ossification centers appear in each mandibular half by the eighth week. One ossification center appears in the lower border, in front of the alveolar process, one in the distal end of Meckel's cartilage in the region of the symphysis, one for the coronoid, one in the cartilage for the condyle and top of ramus, one at the mandibular angle, and one for the inner alveolar plate. They all coalesce together by the twelfth week except for the cartilage-bound condyle that develops into two parts by adolescence, separated by a fibrous strip to help mold it into shape. As the fibrous strip recedes, the condyle fuses to the mandible.

FACIAL ANOMALIES

A-2.1. Facial Clefts

These are very rare, arising as the frontonasal process median and lateral prominences, and maxillary prominences grow toward each other to unite. Ectodermal tissue overlies all of the developing mesenchymal facial parts and forms grooves between them. As the mesenchymal prominences move toward each other, they penetrate the ectodermal tissue in the grooves and unite. Failure of penetration at the critical threshold time of union leaves a fissure between the opposing prominences. Hypoplasia with developmental delay affecting one or both of the adjacent facial parts is often the cause with the resulting cleft formation reflecting the degree of developmental delay (Fig. A-2.1.1). The cleft is usually unilateral, complete or incomplete, and fissures can occur anywhere in the facial region where two parts

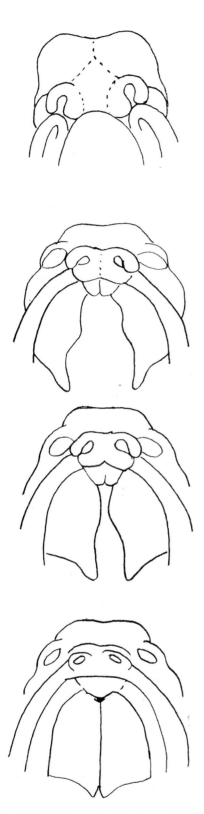

FIGURE A-2.0.2. Embryonic development of the palate: top to bottom—the inferior portion of the two subparts of the frontonasal process unite and extend downward to form the primary palate between the developing maxillary palatine shelves that grow toward each other, uniting with the premaxillary primary palate.

are programmed to meet. With failure of the maxillary and lateral nasal prominences to unite, a nasomaxillary cleft forms between the maxillary and lateral nasal prominence, from the oral cavity to the eye orbit. The naso-ocular groove between the lateral and median nasal prominence becomes the nasolacrimal duct, but when the two parts fail to reach each other at the critical threshold time, the two become divided by a naso-ocular cleft that may extend to the oral cavity (Fig. A-2.1.2). Bilateral clefts can be symmetrical or asymmetrical, and severe forms are generally associated with developmental disturbances of the brain and do not survive fetal life. Mild forms of facial cleft formation often reach adulthood with the cleft usually covered by fibrous tissue (Burdi et al. 1988; Mladick et al. 1974).

Midface hypoplasia of one or both parts of the median nasal prominence creates a median or midline facial cleft as the two parts fail to come together and fuse on time (Figs. A-2.1.1C and A-2.1.2C), often leaving a wide gap between the eye orbits (hypertelorism). The premaxilla may have central notching of the nasal border or a cleft extending between the central incisors. The nares are wide and the nasal root is broad with wide nasal bones attached at odd angles, frontal sinuses are atypical or absent, and the vomer and perpendicular plates of the ethmoid are usually hypoplastic.

Severe median clefts from aplasia or severe hypoplasia are usually associated with brain defects and are very rare and fatal during the perinatal period, with the eye orbits forced close together (hypotelorism) as the bones of the nares and premaxilla fail to develop (DeMyer 1967).

A-2.2. Nasal Bone Hypoplasia/Aplasia

Small or absent nasal bones can occur independent of other facial developmental disturbances resulting in unilateral or bilateral expressions (Fig. A-2.2).

A-2.3.1. Cleft Lip

It is actually a cleft premaxilla, including its primary palate. This usually involves unilateral or bilateral hypoplasia or aplasia of the premaxilla, leaving a separation between adjacent bony structures. Mild forms of clefting appear as notches between affected bony parts. Since the incisors are contained within the premaxilla, wherever the cleft penetrates the alveolar margin, the development of the associated incisor or incisors is disrupted, most often with agenesis. Incomplete expressions are usually expressed as unilateral hypoplasia with some union of the two premaxillary halves. The alveolar margin may be thin and directed upward with central incisors disturbed or absent. The nares on the hypoplastic side may appear lower than the other side. Expressions of cleft lip are varied as clefts can form between the maxilla and premaxilla or between the two

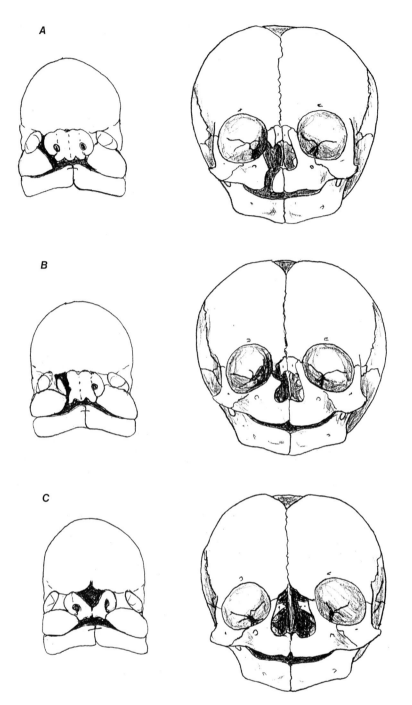

FIGURE A-2.1.1. Facial cleft development: from embryo to newborn—(*A*) nasomaxillary cleft; (*B*) naso-ocular cleft; (*C*) median cleft.

halves of the premaxilla. Unilateral cleft lip forming between the maxilla and premaxilla, most often on the left side, is more common than bilateral expressions, and this type has a greater chance of survival. Severe bilateral cleft with remnant premaxilla usually presents as a rounded ball-like extension between the clefts with all incisors absent or grossly distorted (Fraser 1963). Midline cleft lip from failure of the two subparts of the premaxilla to unite can vary from a slight cleft to a wide cleft with agenesis of one or both halves. Mild expressions of

midline clefts appear as notches or indentations between the central incisors (Fig. A-2.3.1).

A-2.3.2. Cleft Lip with Cleft Palate

The majority of cleft lip (cleft premaxilla) expressions are accompanied by cleft maxillary palate. Premaxilla hypoplasia or aplasia, complete or incomplete, often disrupts the development of the adjacent maxillary palate (Figs. A-2.3.2–A-2.3.6).

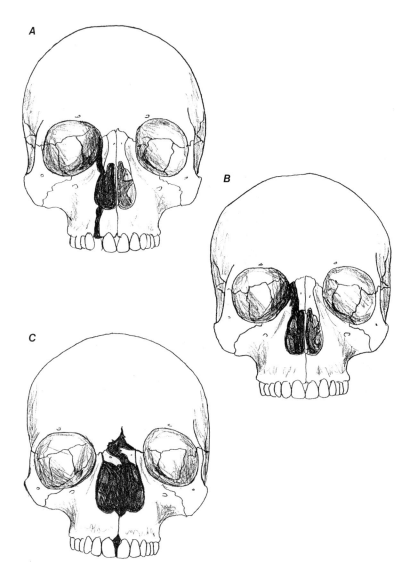

FIGURE A-2.1.2. Facial clefts: (adult) (*A*) nasomaxillary cleft; (*B*) naso-ocular cleft; (*C*) median cleft with wide nares and hypertelorism.

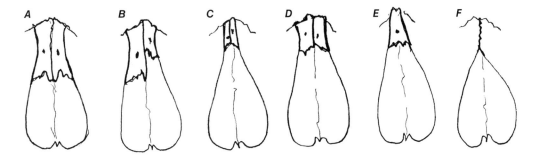

FIGURE A-2.2. Nasal bone hypoplasia/aplasia: (*A*) normal; (*B*) unilateral left hypoplasia; (*C*) bilateral severe hypoplasia; (*D*) bilateral mild hypoplasia; (*E*) single nasal bone from unilateral aplasia; (*F*) bilateral aplasia.

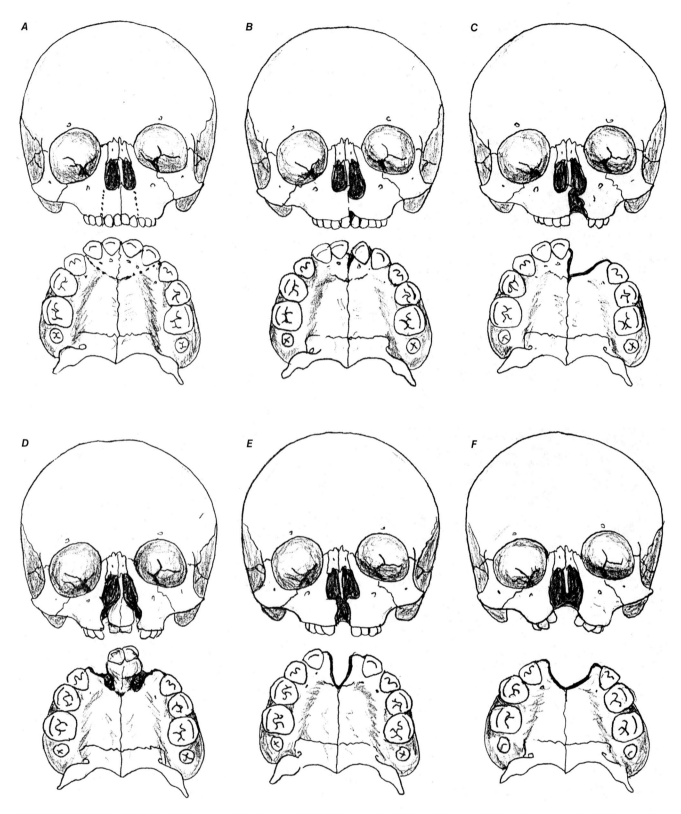

F I G U R E A-2.3.1. Cleft lip (cleft premaxilla): (young child) (*A*) normal with dotted lines outlining the premaxilla; (*B*) incomplete unilateral left cleft; (*C*) complete left unilateral cleft; (*D*) bilateral cleft; (*E*) midline cleft; (*F*) agenesis of the premaxilla—wide cleft.

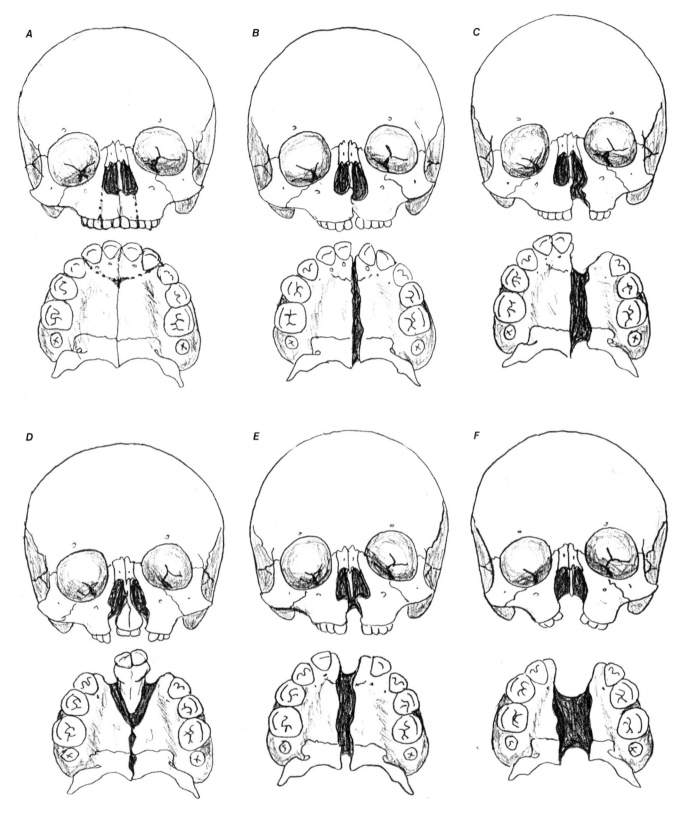

FIGURE A-2.3.2. Cleft lip (premaxilla) with cleft (maxillary) palate: (young child) (*A*) normal with dotted lines outlining the premaxilla; (*B*) incomplete left cleft lip with unilateral left cleft palate; (*C*) unilateral left cleft lip and palate; (*D*) bilateral cleft lip and palate; (*E*) midline cleft lip and palate; (*F*) agenesis of the premaxilla with wide midline cleft palate.

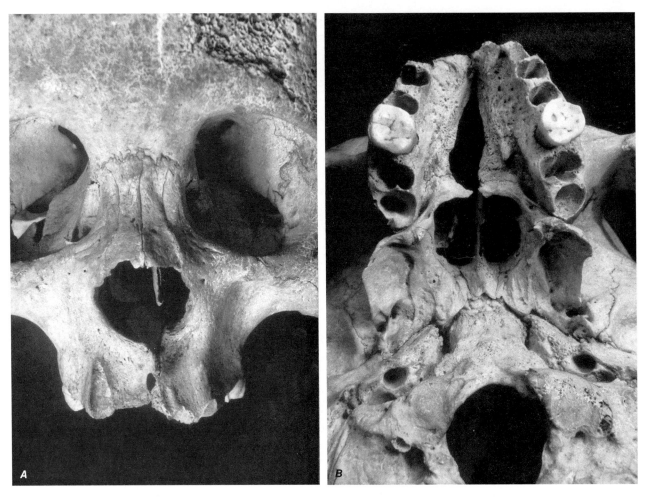

FIGURE A-2.3.3. Cleft lip with cleft palate: (*A*) unilateral right incomplete cleft lip with (*B*) unilateral right cleft palate, adult female (NMNH 316482), SW Colorado.

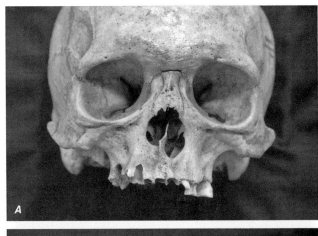

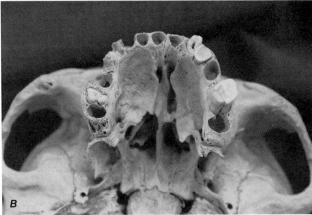

FIGURE A-2.3.4. Cleft lip with cleft palate: (*A*) unilateral left incomplete cleft lip with (*B*) unilateral left cleft palate, adult male (NMNH 266052), Pachacamac, Peru.

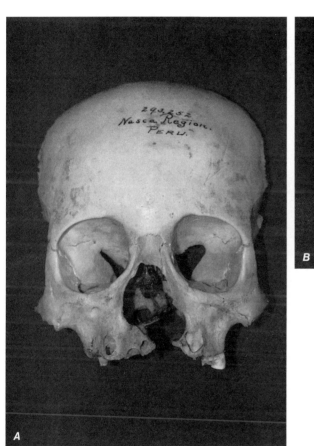

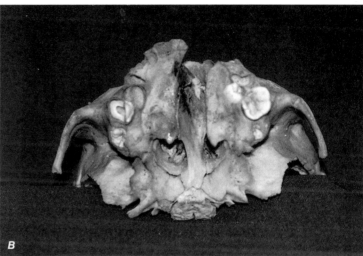

FIGURE A-2.3.5. Cleft lip with cleft palate: (*A*) unilateral left cleft lip with (*B*) unilateral cleft palate, 8- to 10-year-old child (NMNH 293262), Nasca, Peru.

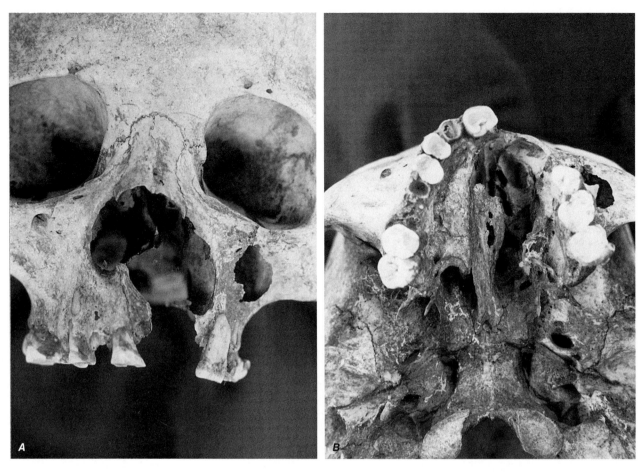

FIGURE A-2.3.6. Cleft lip with cleft palate: (*A*) unilateral left cleft lip with (*B*) unilateral cleft palate, adult female (ASU), Sandoval village, SW Colorado.

A-2.4. Cleft Palate

Isolated cleft maxillary palate evolves as a separate disturbance from cleft lip with cleft palate. Delay in the development and descent of the primitive tongue from the nasal region can interfere with the timing of the approach of the two maxillary palatine shelves toward each other as they meet first with the premaxillary primary palate and end by fusing at the dorsal border. Hypoplasia or aplasia of one or both halves is generally the cause. Cleft palate can be unilateral (the most common form), bilateral asymmetrical, or symmetrical, mild or severe (Fig. A-2.4.1). Mild forms can be expressed as notching of the dorsal border (Fig. A-2.4.2). Bilateral clefts leave the vomer unattached, while a unilateral cleft leaves one side united with the vomer. Clefting also affects the developing soft palate and uvula (Freni and Zapisek 1991).

A-2.5. Cleft Mandible

Very rarely, the two mandibular halves may fail to unite during the first year, with the two halves held together by fibrous tissue (Fig. A-2.5). Mild expressions can

appear as a central notch or indentation between the central incisors (Weinberg and Van de Mark 1972).

A-2.6. Mandibular Hypoplasia

This involves the ascending ramus and dorsal portion of the mandibular body. Unilateral expressions are associated with facial asymmetry known medically as hemifacial microsomia. Bilateral expressions (usually mild) often go undetected unless asymmetrical in expression. Adjacent maxillary bone can be affected, particularly the zygomatic process. Various expressions can range from mild to severe (Fig. A-2.6.1). Type I hemifacial microsomia involves only mild/moderate mandibular hypoplasia, while type II involves a greater degree of hypoplasia with a narrow abnormally shaped ramus (Figs. A-2.6.2–A-2.6.5). Aplasia affecting most of the ramus signifies a type III expression (Kaban et al. 1981).

A-2.7. Bifid Mandibular Condyle

This develops when the fibrous tissue septa acting as scaffolding for the calcifying cartilage extending into the

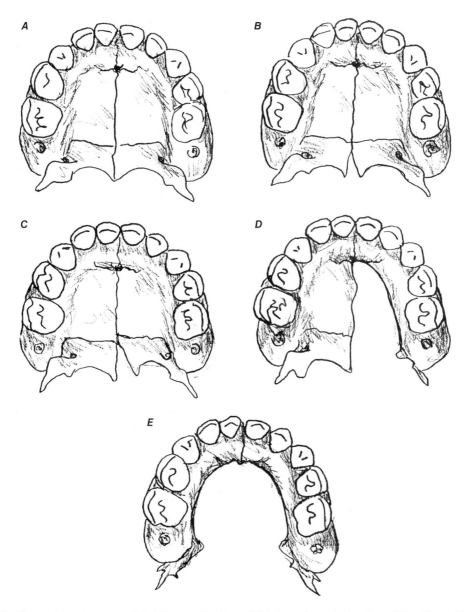

F I G U R E A-2.4.1. Cleft palate: (young child) (*A*) normal palate; (*B*) bilateral notched palate; (*C*) unilateral left notched palate; (*D*) unilateral left cleft; (*E*) complete bilateral cleft.

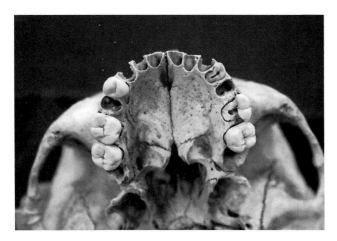

FIGURE A-2.4.2. Bilateral notched cleft palate: adult female (NMNH 264519), Chicama, Peru.

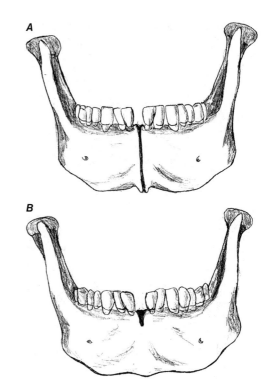

FIGURE A-2.5. Cleft mandible: (*A*) complete cleft; (*B*) notched mandible.

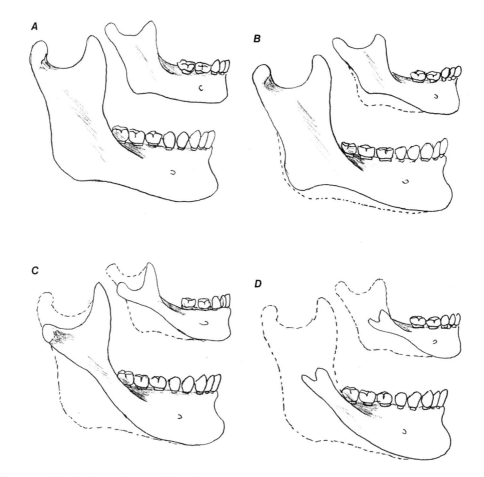

FIGURE A-2.6.1. Mandibular hypoplasia: (adult and child) (*A*) normal; (*B*) type I; (*C*) type II; (*D*) severe type III (dotted lines represent normal).

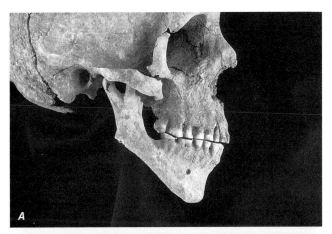

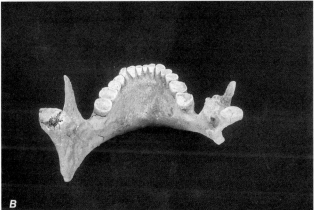

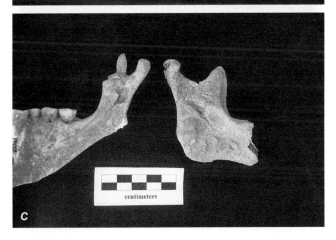

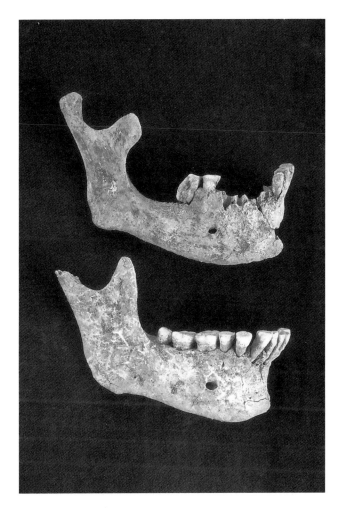

FIGURE A-2.6.3. Mandibular hypoplasia: unilateral right, adult female mandible compared with (lower) normal mandible, La Playa, NW Mexico.

FIGURE A-2.6.2. Mandibular hypoplasia: unilateral right; (*A*) right side of the face, (*B*) dorsal view of the mandible, and (*C*) right and left sides of the mandible, adult male, La Playa, NW Mexico.

mandibular head of the ramus fails to recede (Fig. A-2.7.1). Postnatal bifurcation can also occur with injury to the condylar head. Whereas the trauma-induced bifurcation is oriented anteroposteriorly, developmental bifurcation is positioned mediolaterally (Fig. A-2.7.2). Bifid condyles have a corresponding bifid articular temporal joint space. Unilateral expressions are more

common than bilateral expressions (Blackwood 1957; McCormick et al. 1989).

A-2.8. Coronoid Hyperplasia

This is a genetically programmed progressive hyperdevelopment from the coronoid ossification center, culminating in enlarged bilateral coronoids by adolescence (Figs. A-2.8.1 and A-2.8.2). This prevents the mouth from fully opening as the expanded coronoid processes impinge on the posterior aspects of the zygomatics. It occurs mostly in males from the same family (Schultz and Theisen 1989).

A-2.9. Palate Inclusion (Fissural) Cyst

Overlying ectodermal tissue of the developing palate fails to retreat as membranous bones of the palate come together, leaving a pocket filled with fluid or semisolid

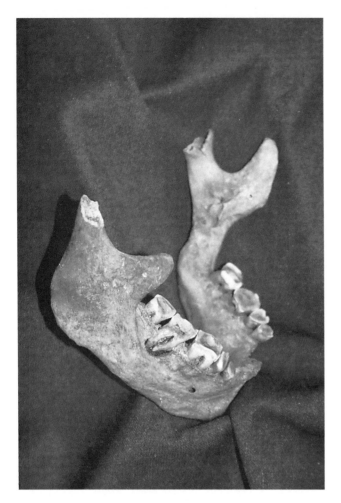

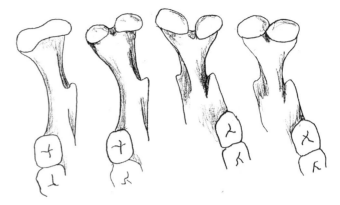

FIGURE A-2.7.1. Bifid mandibular condyle: normal (left) with examples of developmental bifurcation.

FIGURE A-2.6.4. Mandibular hypoplasia: unilateral left, adult female mandible (NMNH 242146), Dos Pueblos site, Santa Cruz Island, CA.

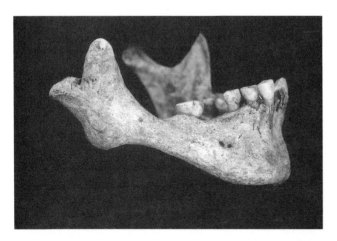

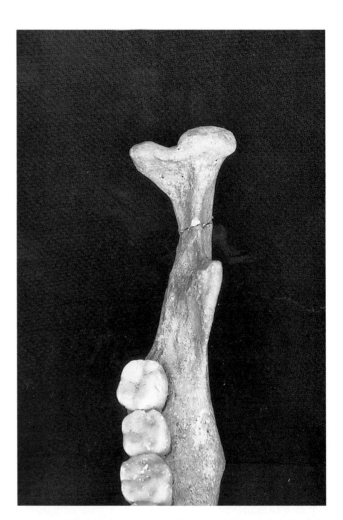

FIGURE A-2.6.5. Mandibular hypoplasia: unilateral right, adult female mandible, Irene Mound, Mississippian Mound Complex (contributed by Clark Spencer Larsen).

FIGURE A-2.7.2. Bifid mandibular condyle: left side of the bilateral expression, adult female, Byzantine Petras, Crete.

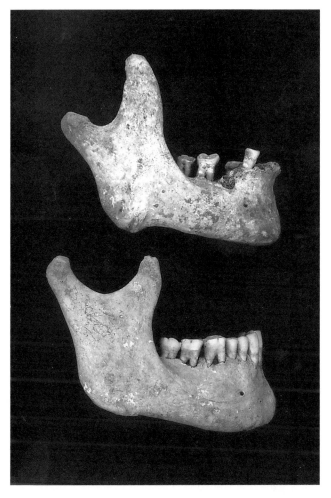

FIGURE A-2.8.1. Mandibular coronoid hyperplasia: (top) affected adult male compared with (bottom) normal adult male, Frankish Corinth, Greece.

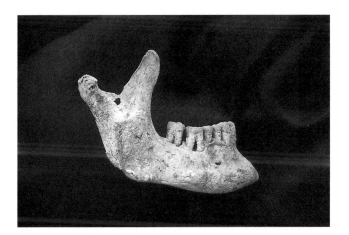

FIGURE A-2.8.2. Mandibular coronoid hyperplasia: adult male, Classical Period, Corinth, Greece.

material bounded by trapped epithelial cells within the developing bone. The incisive canal marks the junction of the premaxillary primary palate with the paired maxillary palatine shelves and is the most common site for development of an inclusion cyst—the median anterior inclusion cyst (Figs. A-2.9.1A and A-2.9.2A). The median palatal inclusion cyst develops midline between the two palatine shelves, and it can grow quite large over time (Fig. A-2.9.1B). The globulomaxillary inclusion cyst forms at the junction of the premaxilla primary palate and one of the maxillary palatine shelves (Fig. A-2.9.1C), between the roots of the lateral incisor and canine teeth (Little and Jakobsen 1973; Schafner et al. 1983).

A-2.10. Mandibular Inclusion Cyst

This is commonly referred to as the Stafne defect and takes shape as part of the primordial sublingual salivary gland bordering the submandibular fossa of the membranous bone develops prematurely, impinging upon developing membranous bone (Figs. A-2.10.1 and A-2.10.2). As the gland expands with age, it leaves a shallow or deep oval depression below the mylohyoid line near the inferior border of the inner retromolar aspect of the mandibular body (Stafne 1942; Wolf et al. 1986).

A-2.11. Mandibular Torus

This is a form of hyperplasia developing from the inner alveolar plate ossification center, resulting in one or more rounded bony protuberances below the alveolar margin on the lingual side of the mandible, generally between the canine and first molar. This is most often bilateral with more than one protuberance, varying in size and shape on each side. Sometimes the torus develops unilaterally, and they can appear as mild expressions to quite large in form (Figs. A-2.11.1–A-2.11.3).

Bony hyperplasia can also form along the median suture of the palate to form a palatine torus, but it is usually much less significant. Early development of both types of torus has been noted in infancy and early childhood, enlarging with maturity (Hauser and De Stefano 1989).

A-3. EXTERNAL AUDITORY MEATUS AND TYMPANIC PLATE DEVELOPMENT

While the first pharyngeal arch contributes to facial bones, the distal end of the ectodermal groove separating the external arch from its internal endothelial lined pouch leads to the development of the external auditory

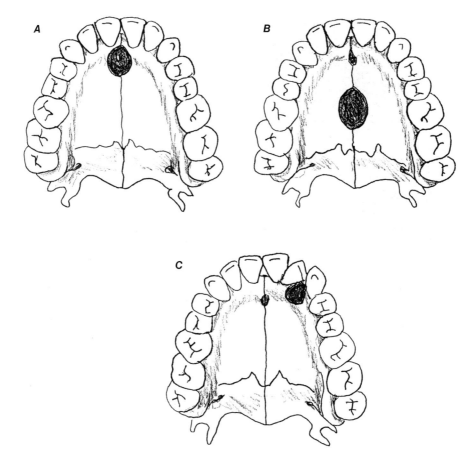

F I G U R E A-2.9.1. Palatal inclusion (fissural) cysts: (*A*) median anterior within incisive canal; (*B*) median palatal between palatal plates; (*C*) globulomaxillary at the junction of premaxilla and maxilla.

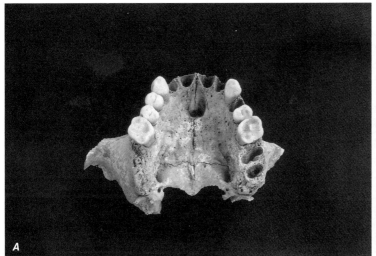

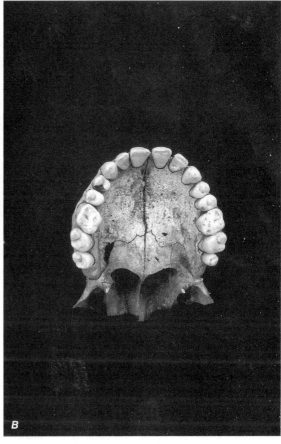

FIGURE A-2.9.2. Palatal inclusion (fissural) cyst: (*A*) adult median anterior cyst within the incisive canal, compared with (*B*) adult palate normal incisive canal (note left peg third molar), Frankish Corinth, Greece.

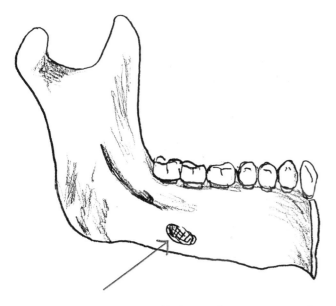

FIGURE A-2.10.1. Mandibular inclusion cyst (Stafne defect).

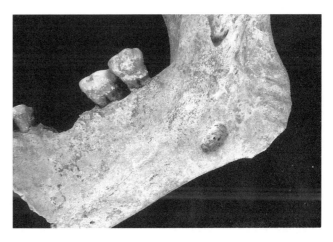

FIGURE A-2.10.2. Mandibular inclusion cyst (Stafne defect): unilateral right, adult male, La Playa, NW Mexico.

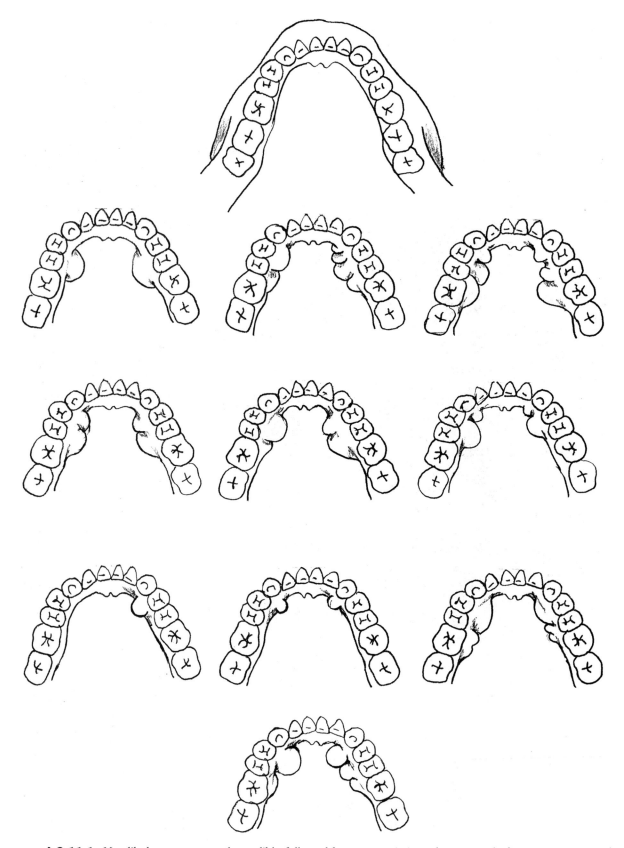

FIGURE A-2.11.1. Mandibular torus: normal mandible followed by torus variations drawn mostly from American Southwest skeletal collections.

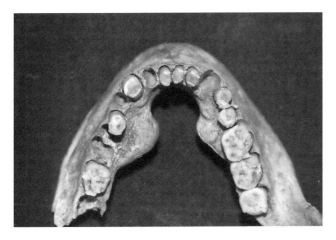

F I G U R E A-2.11.2. Mandibular torus: adult male (NMNH 262945), Puye, NM.

meatus (Fig. A-3.0). The distal end of the endodermal pouch grows sack-like to form the inner ear's tympanic cavity while the proximal end remains narrow to form the eustachian tube as the external anterior portion of the ectodermal groove disappears.

The thin membrane separating the pouch and groove forms the membranous tympanic plate—the floor of the external auditory meatus—as the distal end of the ectodermal groove creates the external ear opening. By the ninth week, four small ossification centers form around the edges of the tympanic membrane, coalescing to form a U-shaped tympanic ring that eventually fuses to the squamosal part of the temporal bone by the thirty-fifth week. At birth, the bony tympanic ring frames the fibrocartilaginous tympanic plate that grows with the expansion of the auditory canal.

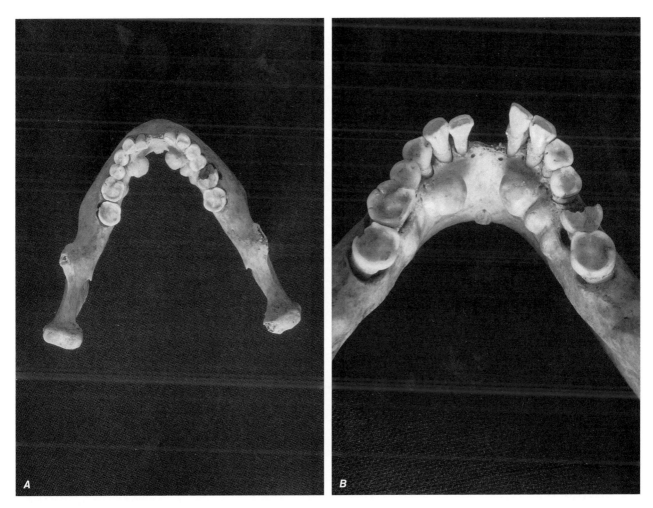

F I G U R E A-2.11.3. Mandibular torus: (A) adult female, (B) close-up, Bronze Age Da Shan Qian, Inner Mongolia, PRC.

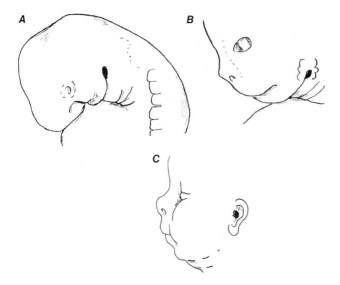

FIGURE A-3.0. External auditory meatus development: (*A*) fifth embryonic week; (*B*) seventh embryonic week; (*C*) newborn.

Ossification spreads throughout the tympanic plate rapidly from the anterior and posterior portions, but they are slow to merge, leaving an unossified area that appears as an opening in the bony tympanic plate, often referred to as foramen of Huschke. But it is not a true foramen as it essentially remains fibrocartilage until final ossifying closure by age 5 (Fig. A-3.2A–D). Bony extensions from the tympanic plate form a sheath around the developing styloid and the outer border juts out in semifolds from the external auditory meatus.

EXTERNAL AUDITORY MEATUS AND TYMPANIC PLATE ANOMALIES

A-3.1. Atresia (Aplasia)/Hypoplasia External Auditory Meatus

Complete aplasia is a rare occurrence when the dorsal end of the first ectodermal groove fails to develop, usually unilaterally with the right side affected more than the left. The tympanic plate does not form and the styloid is rudimentary or absent and the petrous portion may be smaller (Hrdlicka 1933). The external soft tissue auricle of the ear is also affected. Hypoplasia can also occur with the development of the external auditory meatus producing a narrow opening with smaller tympanic plate (Figs. A-3.1.1–A-3.1.3).

A-3.2. Tympanic Aperture

When the tympanic plate fails to completely ossify from its fibrocartilaginous precursor, it leaves the bony

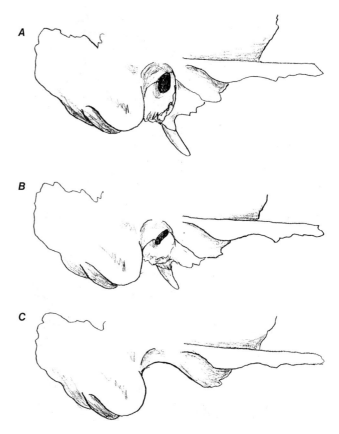

FIGURE A-3.1.1. Atresia (aplasia)/hypoplasia external auditory meatus: (*A*) normal opening; (*B*) narrow opening caused by hypoplasia; (*C*) atresia (aplasia).

opening to persist throughout life. This aperture can vary in size, and occasionally, the fibrocartilaginous tympanic plate fails completely to ossify, leaving a wide cleft in the floor of the external auditory meatus (Fig. A-3.2E,F).

A-3.3. External Auditory Torus

This is often referred to as external auditory exostosis but differing from pathological bony tumors (Hutchinson et al. 1997; Mann 1984). The torus forms with growth at the junction border between the tympanic plate and the temporal squamosal part of the auditory canal as a smooth bony ripple or nodule that can vary in size from slight to a large, usually as a single expression, sometimes double. Hyperplasia of one or both nodules on the primordial arms of the tympanic ring leads to this development that usually appears bilateral. Most often, it occurs at the inferior junction but can develop at the superior junction or both (Figs. A-3.3.1 and A-3.3.2).

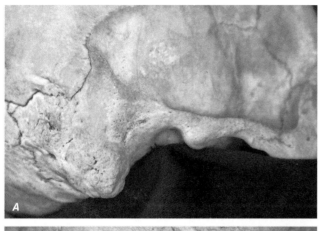

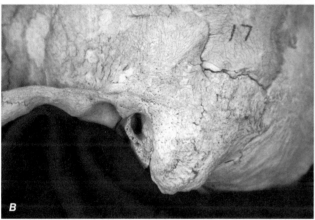

F I G U R E A-3.1.2. Atresia (aplasia) external auditory meatus: (*A*) unilateral right, (*B*) normal left side, adult male (NMNH 264542), Chicama, Peru.

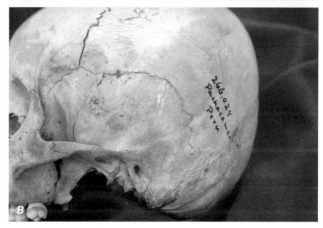

F I G U R E A-3.1.3. Atresia (aplasia) external auditory meatus: (*A*) unilateral right, (*B*) normal left side, child (NMNH 266024), Pachacamac, Peru.

A-4. STYLOHYOID CHAIN DEVELOPMENT

The remaining pharyngeal arches contribute to the cartilages in the neck, while the associated pouches evolve into various glands as the ectodermal grooves between them disappear. The laryngeal cartilages—thyroid, cricoid, arytenoid, corniculate, and cuneiform cartilages—evolve from the last two pharyngeal arches, while the second and third arches provide cartilage for the stylohyoid chain. As cartilage tissue never looses the potential to ossify, sometimes trauma or genetic programming can lead to ossification of the laryngeal cartilages. Ossification of adult male thyroid cartilage is not unusual.

Paired Reichert's cartilages from the second pharyngeal arch form the bony styloid processes, stylohyoid ligaments connecting the styloids to the lesser cornua of the hyoid, and form the upper body of the hyoid as the ends of the cartilages meet midline. Paired cartilages from the third arch provide the greater cornua and lower part of the hyoid body. The dorsal end of

Reichert's cartilage separates and becomes enclosed in the tympanic cavity to form the stapes.

The styloid chain begins with the proximal base of the styloid (tympanohyal) ensheathed by the tympanic plate. The apex of the elongated distal portion of the styloid (stylohyal) forms the attachment for the stylohyoid ligament (epihyal) that connects with the lesser cornua (hypohyal) attached to the superior portion of the hyoid at its junction with the greater cornua. The tympanohyal and stylohyal ossify from separate ossification centers, with the former present at birth and the latter appearing after birth, eventually uniting after puberty (Fig. A-4.0).

STYLOHYOID CHAIN ANOMALIES

A-4.1. Stylohyoid Chain Variations in Ossification

The stylohyal may not ossify leaving a shortened, hypoplastic (rudimentary) styloid (Fig. A-4.1.1B). Both

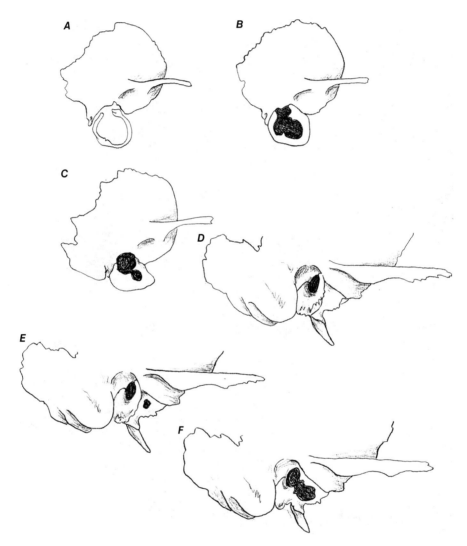

FIGURE A-3.2. Tympanic aperture: (*A*) tympanic ring at birth; (*B*) ossifying tympanic ring at 1 year; (*C*) continued ossification at 2 years; (*D*) adult complete ossification; (*E*) adult tympanic aperture; (*F*) adult tympanic cleft.

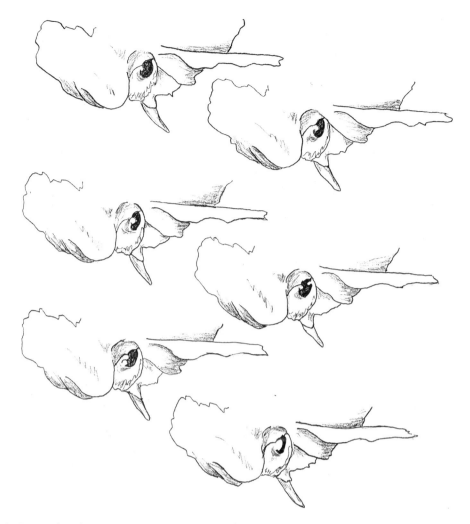

F I G U R E A-3.3.1. External auditory meatus torus: variations from La Playa, NW Mexico.

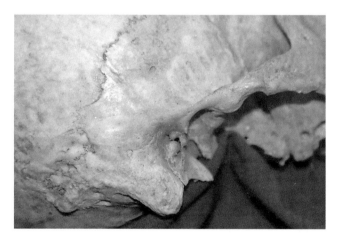

F I G U R E A-3.3.2. External auditory meatus torus: adult male (NMNH 266023), Pachacamac, Peru.

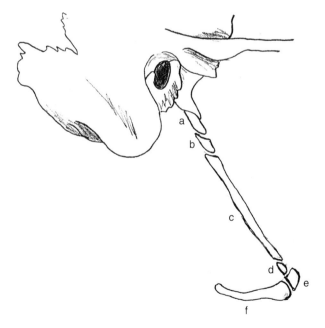

F I G U R E A-4.0. Stylohyoid chain segments: (a) proximal end of the bony styloid process—tympanohyal; (b) bony distal end of the styloid process—stylohyal; (c) stylohyoid ligament—epihyal; (d) lesser cornu of the hyoid—hypohyal; (e) hyoid body; (f) greater cornua of the hyoid.

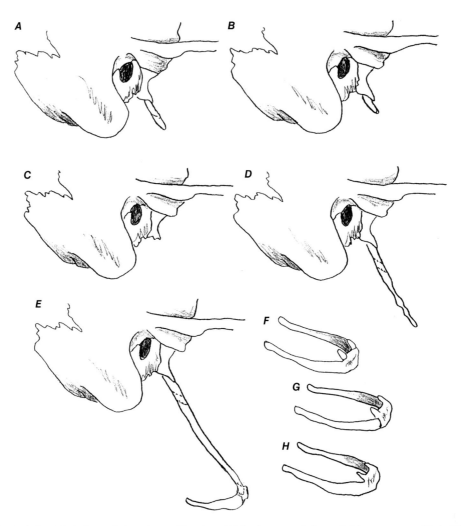

FIGURE A-4.1.1. Stylohyoid chain variations in ossification: (*A*) both parts of the styloid process ossified; (*B*) only the proximal portion of the styloid process ossified (bony hypoplasia); (*C*) neither part of the styloid process ossified (bony aplasia); (*D*) ossification of both parts of the styloid process and stylohyoid ligament (bony hyperplasia); (*E*) complete ossification of the stylohyoid chain; (*F*) separate ossified lesser cornua with greater cornua united with the hyoid body; (*G*) lesser cornua ossified and united to the hyoid body; (*H*) both ossified lesser cornua and greater cornua united with the hyoid body.

the stylohyal and tympanohyal may not ossify, leaving the bony styloid absent (Fig. A-4.1.1C). Part or all of the epihyal—the stylohyoid ligament—may ossify with the styloid to produce a greatly elongated bony styloid that may be crooked with a blunt tip (Figs. A-4.1.1D and A-4.1.2), appearing as a hyperplastic styloid process, either unilaterally or bilateral with or without asymmetry. The lesser cornua, attached to the hyoid by fibrous joints, may remain cartilaginous or ossify and sometimes uniting with the hyoid. The bony greater cornua are usually connected by fibrous joints to the hyoid body but may unite with it (Figs. A-4.1.1F and A-4.1.3) if the joint fails to develop. Rarely, both lesser and greater cornua ossify as part of the hyoid body (Figs. A-4.1.1H and A-4.1.4). Very rarely, the entire stylohyoid chain ossifies to unite all of its parts, or each segment ossifies

separately with fibrous connective tissue holding them all together (Camarda et al. 1989; Gossman and Tarsitano 1977).

A-4.2. Thyroglossal Developmental Cyst

This can form anywhere along the descending pathway of the primordial thyroglossal duct as it passes from the base of the tongue to the anterior midline of the throat, past the hyoid body to its final position below the thyroid and cricoid cartilages to form the thyroid gland (Fig. A-4.2.1). Remnant thyroglossal duct tissue left behind forms a cyst that can impress upon adjacent structures such as the hyoid body (Figs. A-4.2.1B and A-4.2.2) (Moore 1985:1027; Sadler 2006:271–272).

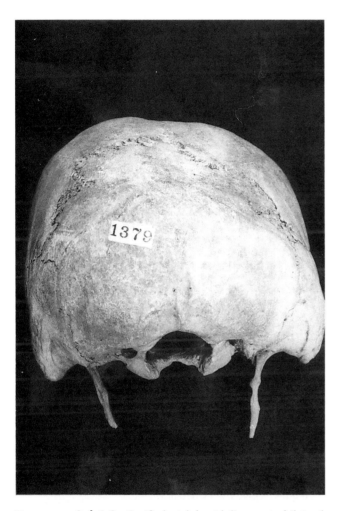

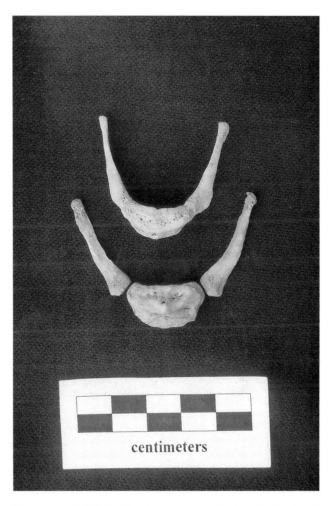

FIGURE A-4.1.2. Ossified stylohyoid ligament: bilateral, united with the bony styloid processes, adult male, Old Walpi, AZ (Field Museum).

FIGURE A-4.1.3. Greater cornua united with the hyoid body: (top) adult female compared with (bottom) the normal separated cornua, adult male, Byzantine Panakton, Greece.

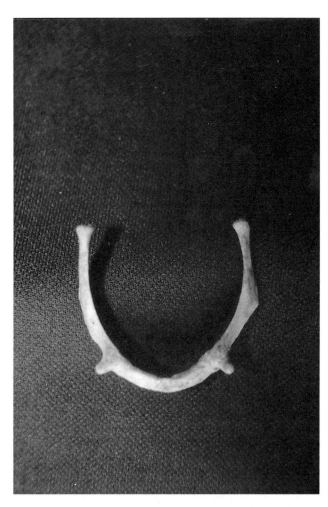

Figure A-4.1.4. Ossified lesser cornua and greater cornua united with the hyoid body: adult male (NMNH 308623), Hawikuh, NM.

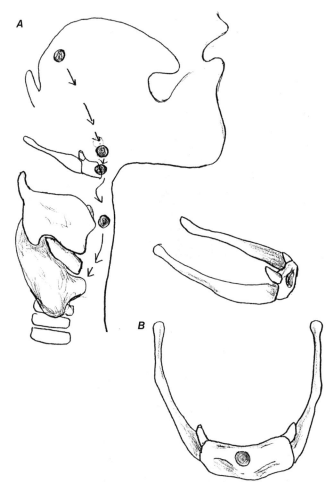

Figure A-4.2.1. Thyroglossal cyst development: (*A*) embryonic descent of primordial thyroglossal duct from below the tongue, past the hyoid body to the final position below the thyroid and cricoid cartilages—circles represent where cysts may form from remnant tissue; (*B*) hyoid body with imprint of the thyroglossal cyst.

FIGURE A-4.2.2. Thyroglossal cyst: (top) normal hyoid body compared with (bottom) the hyoid with thyroglossal cyst, young adult male, Frankish Corinth, Greece.

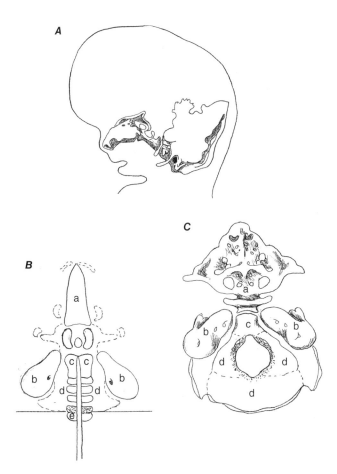

FIGURE A-5.0.1. Skull base (chondocranium) development: (*A*) the developing fetal cartilaginous plate; (*B*) developing precursor cartilages; (*C*) corresponding later development with broken lines representing contributing cranial portion of the first cervical sclerotome (a) ventral trabecular cartilages, (b) lateral otic capsules, (c) post-hypophyseal fossa parachordal cartilages surrounding the cranial end of the neural tube, (d) occipital sclerotmes, and (e) first cervical sclerotome with dotted cranial portion joining occipital ones, line divides occipital base with the atlas.

A-5. SKULL BASE DEVELOPMENT

The forerunner of the skull base—the chondocranium—forms by the end of the first month from a cartilaginous plate known as the prechordal cranial base, extending from the nasal region to surround the cranial end of the neural tube, thus cradling the developing brain. This plate stems from the expansion and fusion of three basic pairs of cartilage precursors. The ventral cartilages (trabecular cartilages) developing in the interorbital–nasal region to ultimately form the nasal cartilages, ethmoid, lesser wings, and roots of greater wings as well as the body of the sphenoid. Lateral cartilages (otic capsules) develop around the otocysts, giving rise to the petromastoids of the temporals. Paired cartilages (parachordal cartilages) just distal to the fossa for the hypophysis join together and develop into the basioccipital with small ventral portions of the occipital

condyles. Abutting the parachordal cartilages, four somitomeres condense into occipital somites to form three nonsegmenting sclerotomes. The occipital sclerotomes fuse with the parachordal cartilages, while the cranial portion of the resegmenting first cervical sclerotome splits from its caudal portion to join the last occipital sclerotome in the formation of the lateral exoccipitals and supraoccipital bones (Fig. A-5.0.1). The supraoccipital quickly fuses with the occipital squamosa at the highest nuchal line (the mendosa line). The lateral exoccipitals unite with the supraoccipital by the third year and with the basioccipital by the sixth year (Fig. A-5.0.2).

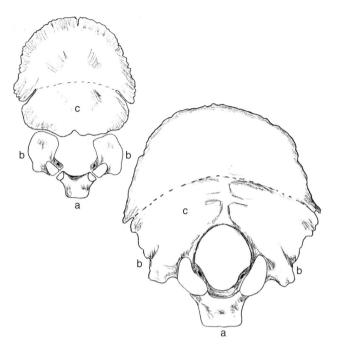

FIGURE A-5.0.2. Skull base (chondocranium): (infant and adult) (a) basioccipital; (b) lateral exoccipitals; (c) supraoccipital, fusion with occipital squamosa represented by broken lines.

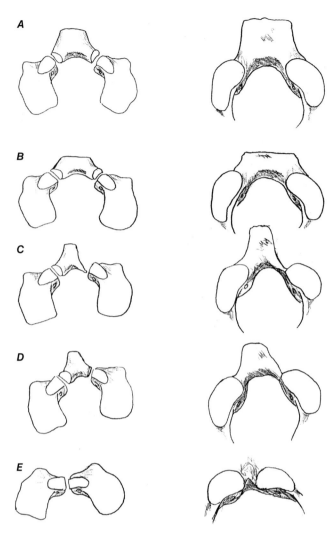

FIGURE A-5.1. Basioccipital hypoplasia/aplasia: (infant and adult) (*A*) normal basioccipital; (*B*) bilateral hypoplasia; (*C*) unilateral left aplasia; (*D*) unilateral left hypoplasia; (*E*) bilateral aplasia.

SKULL BASE ANOMALIES

A-5.1. Basioccipital Hypoplasia/Aplasia

Developmental disturbance of one or both parachordal cartilages can result in hypoplasia or aplasia of the basioccipital with or without distortion of the foramen magnum (Fig. A-5.1).

A-5.2. Basioccipital Clefts

Failure of the two parachordal cartilages to completely fuse presents as a vertical cleft (Fig. A-5.2A). Cranial shifting of the border between the atlas and occipital that moves the border upward can be variably expressed, and on rare occasions, the shift affects the developing parachordal cartilages, producing bilateral or unilateral horizontal clefts. The nonpathological clefts are filled with cartilaginous fibers in life. This kind of shift can also delete one or both ventral portions of the occipital condyles forming on the dorsal ends of the parachordal cartilages resulting in the appearance of condylar hypoplasia (Fig. A-5.2B–D).

OCCIPITAL–CERVICAL (O-C) BORDER DEVELOPMENT

The O-C border is between the occipital base, exoccipitals, axis apical dens and the atlas, axis (Fig. A-5.3.0). Note that the apical dens develops above the border in the primordial region of the proatlas of early life forms before descending below the border to join the rest of the axis as the permanent border develops (Shapiro and Robinson 1976). See the development of vertebral column under Chapter B for more details.

A-5.3. Cranial Shifting of the O-C Border

The border between the base of the occipital and atlas is moved upward (Fig. A-5.3.1), affecting the last

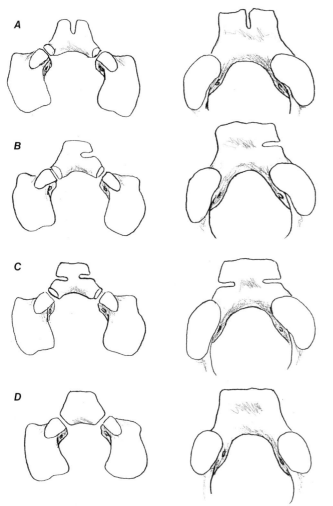

FIGURE A-5.2. Basioccipital clefts: (infant and adult) (*A*) vertical cleft from failure of parachordal cartridges to completely fuse; (*B*) unilateral left horizontal cleft caused by attempted cranial border shift; (*C*) bilateral horizontal cleft from attempted cranial border shift; (*D*) bilateral agenesis of ventral portions of occipital condyles from attempted cranial border shift (hypoplasia of occipital condyles).

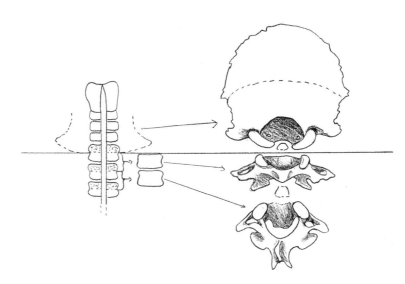

FIGURE A-5.3.0. Occipital–cervical border: represented by a solid line with basioccipital and apical dens forming above the O-C border, atlas, dens, and axis body developing below the O-C border.

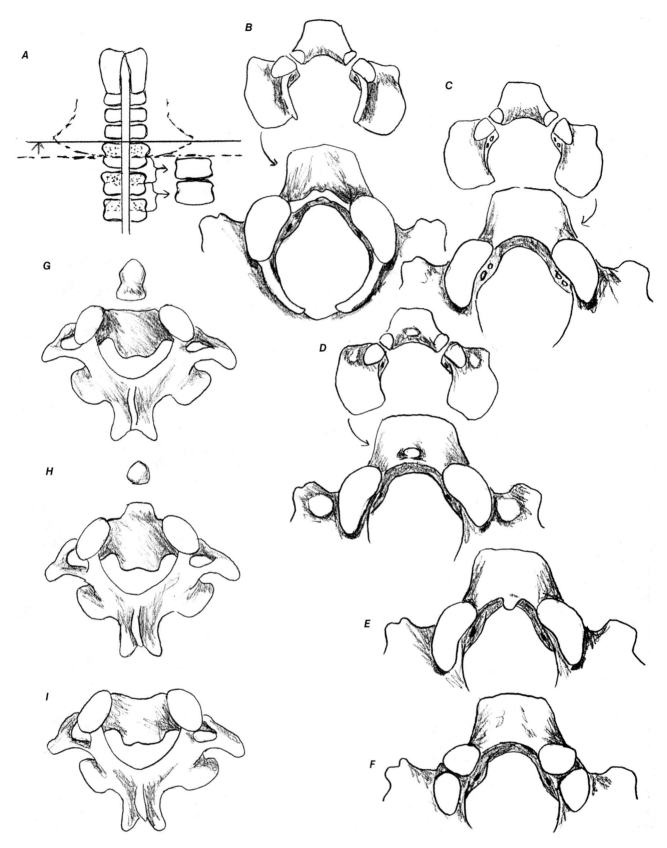

F I G U R E   A-5.3.1. Cranial shifts at the occipital–cervical (O-C) border: (*A*) schematic drawing shows the border (solid line) moving upward from the normal border represented by a dashed line; (newborn and adult) basioccipital; (*B*) complete shift occipital vertebra expression; (*C*) mild shift double hypoglossal canals; (*D*) precondylar and paracondylar (bony protrusions) processes; (*E*) precondylar extended tubercle from the anterior rim of the foramen magnum; (*F*) bifurcated occipital condyles; (*G*) separate dens (type I dens defect—os odontoideum); (*H*) separate dens' tip (type II dens defect—ossiculum terminale); (*I*) agenesis of dens, both apical tip and base (type V dens defect).

FIGURE A-5.3.2. O-C cranial border shift precondylar tubercle: adult female and adult male, Warring States Period, Da Shan Qian, Inner Mongolia, PRC.

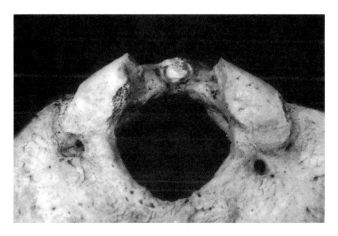

FIGURE A-5.3.3. O-C cranial border shift separated the apical dens' tip attached to the foramen magnum rim: adult male (NMNH 271905), Amoxiumqua, NM.

FIGURE A-5.3.4. O-C cranial border shift precondylar process: adult male (NMNH 308602), Hawikuh, NM.

occipital sclerotome as it coalesces with the cranial portion of the resegmenting first cervical sclerotome (Fig. A-5.3.1A). The upward shift forces this segment into attempted separation from the base of the skull to form a substitute proatlas or occipital vertebral segment. Complete separation from the base of the skull is not possible, generally leaving only variable expressions of an occipital vertebra arising from the exoccipitals, represented by attempts to form vertebral arching around the foramen magnum. Usually, there is some form of anterior arch with an incomplete posterior arch protruding around the foramen magnum that often appears distorted by the unusual projections.

Minor forms of cranial border shifting include a small bony tubercle extending from the anterior rim of the foramen magnum as if simulating a pseudo-apical dens tip (Figs. A-5.3.1E and A-5.3.2). Raised bony distor-

tions around the foramen magnum may appear, including the precondylar process (sometimes double in form), and other forms of raised bony processes in the paracondylar spaces, bilaterally or unilaterally (Figs. A-5.3.1D, A-5.3.4, and A-5.3.5). Paramastoid processes near the notch for the jugular foramen may also be part of this phenomenon. These bony processes can be small, or large enough to articulate with the atlas. The hypoglossal canals commonly bifurcate with mild cranial border shifts (Fig. A-5.3.1C), and the occipital condyles may also bifurcate (Fig. A-5.3.1F).

Cranial shifting at the O-C border can sometimes, though rarely, affect the cartilaginous precursor apical dens (odontoid) of the axis vertebra developing within the region of the primordial foramen magnum (Dawson and Smith 1979). With a stable developing O-C border, the dens' apical tip descends from the area of the anterior rim of the foramen magnum to unite with the base of the dens as it moves downward to join the body of the axis vertebra (Fig. A-5.3.0). Shifting upward of the

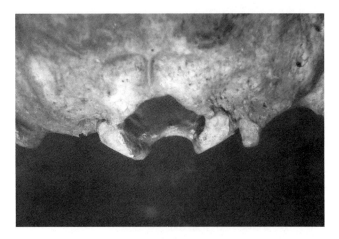

FIGURE A-5.3.5. O-C cranial border shift paracondylar process: adult female (NMNH 271804), Amoxiumqua, NM.

O-C border can delay and prevent the apical tip of the dens from uniting with its base (type II dens defect: ossiculum terminale), leaving it a separate ossicle (Figs. A-5.3.1H and A-5.3.6) that may attach to the anterior of the atlas or anterior rim of the foramen magnum (Fig. A-5.3.3). The entire dens may not join with the axis body (type I dens defect: os odontoideum) (Fig. A-5.3.1G). Agenesis of the tip (type IV dens defect) leaves the dens appearing short and blunt (Fig. A-5.3.7). The base of the dens may fail to ossify within the precursor cartilage, leaving the apical segment as a separated ossicle (type III dens defect) or the entire dens, both base and apical tip, may fail to ossify (type V dens defect) (Fig. A-5.3.1I). With agenesis of the dens or dens apical tip, there will be no articulating facet for it on the atlas (Fig. A-5.3.7). Severe dens defects can lead to serious neurological problems and often involve

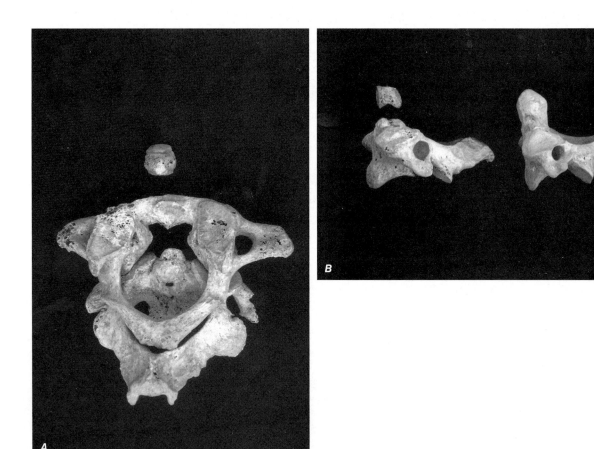

FIGURE A-5.3.6. O-C cranial border shift separated the apical dens' tip: (*A*) atlas with facet for dens and axis with separated apical dens' tip; (*B*) separated apical dens' tip compared with the normal axis dens, adult male, Frankish Corinth, Greece.

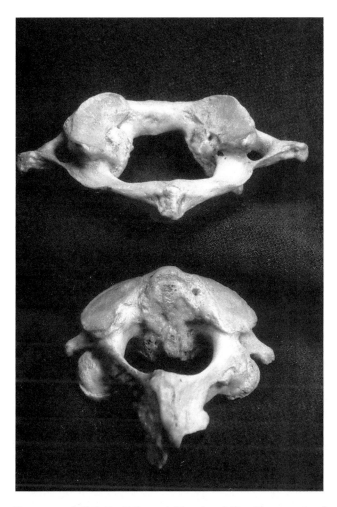

FIGURE A-5.3.7. O-C cranial border shift with agenesis of the apical dens' tip: atlas without facet for dens and axis with agenesis of the apical dens' tip, adult male (778), NMNH Terry collection.

basilar impression—indentation of the base of the skull. Neurological symptoms are often triggered by trauma to the neck that produces subluxation of the O-C (atlanto–axial) junction.

A-5.4. Caudal Shifting of the O-C Border

As the border moves downward between the atlas and axis vertebrae (Fig. A-5.4.1), the atlas becomes distorted as it is assimilated into the skull base (Figs. A-5.4.1B and A-5.4.2). Caudal shifting of the O-C border is far more common than cranial shifting with bilateral or unilateral expressions. The atlas tries to become part of the exoccipitals but fails to loose its vertebral identity. It often displays transverse foramina, whereas the occipital vertebra never does. The anterior arch is usually complete, but the posterior arch is often incomplete and distorted. Distortion of the foramen magnum is often present.

Mild expressions include condylar hypoplasia (Fig. A-5.4.1C) when caudal shifting attempts to obliterate the development of the major portion of the articulating condyles on the exoccipitals. Another form of caudal shifting of this border can affect the apical tip of the axis' dens, forcing it to articulate with the anterior rim of the foramen magnum. A precondylar facet appears on the anterior rim of the foramen magnum with a matching facet on the dens' apical tip (Fig. A-5.4.1D). The adjacent superior anterior arch of the atlas can also be affected with an additional matching facet (Fig. A-5.4.3). Caudal shifting interferes with the timing of the descent of the apical tip of the dens during morphogenesis, allowing it to protrude onto the anterior rim of the foramen magnum.

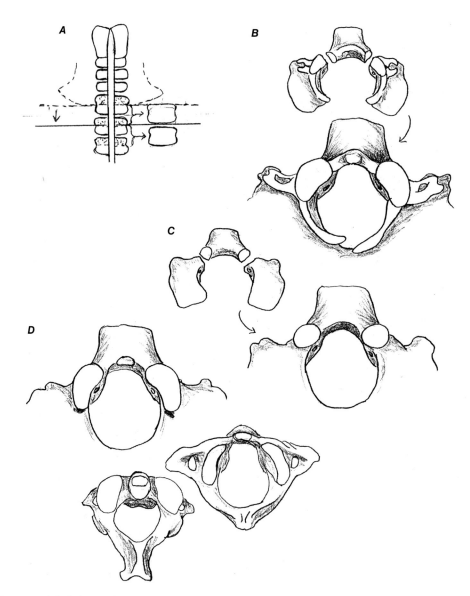

F IGURE A-5.4.1. Caudal shifts at the occipital–cervical (O-C) border: (*A*) schematic drawing shows the border (solid line) moving downward from the normal border represented by a dashed line; (newborn and adult) basioccipital; (*B*) complete shift occipitalized atlas; (*C*) mild shift condylar hypoplasia; (*D*) greater shift with matching articular facets on the anterior rim of the basioccipital, atlas, and dens of axis (not seen on newborn).

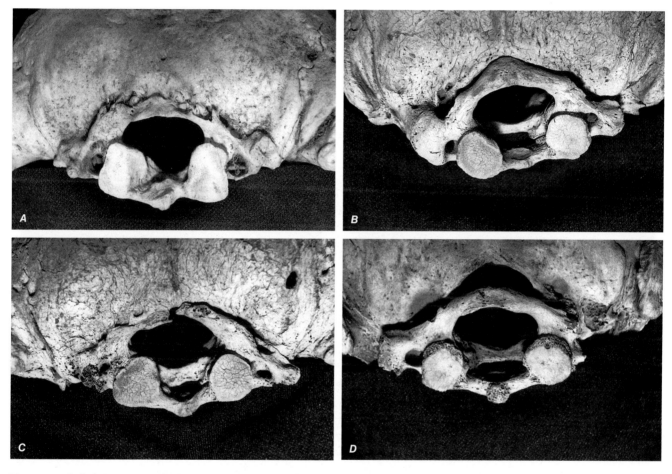

FIGURE A-5.4.2. O-C caudal border shift occipitalized atlas: (*A*) adult male (NMNH 157693), Chaviz Pass, AZ; (*B*) adult female (NMNH 264644), Chicama, Peru; (*C*) adult female (NMNH 264615), Chicama, Peru; (*D*) adult male (NMNH 266361), Pachacamac, Peru.

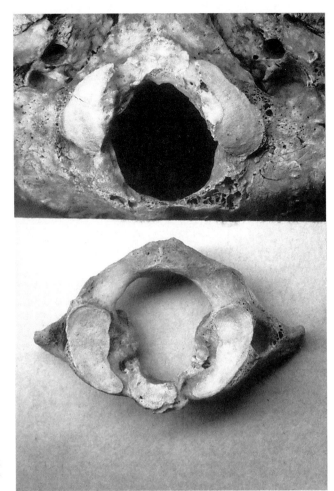

FIGURE A-5.4.3. O-C caudal border shift facets: matching articulating facets anterior rim of the foramen magnum and anterior arch of the atlas, adult female (NMNH 269267), Puye, NM.

CHAPTER B

VERTEBRAL COLUMN

VERTEBRAL COLUMN DEVELOPMENT

The bony vertebral segments derive from the paired mesenchymal tissue columns—paraxial mesoderm that forms along each side of the notochord as it extends below the developing skull base. Whorl-like formations within these tissue columns quickly segment into paired blocks of tissue known as somites. The somites differentiate into ventral–medial sclerotome, lateral and superficial dermatome, and deeper lateral myotome tissue. While myotome and dermatome tissues form skeletal muscle and skin associated with the vertebral segments, sclerotome tissues form the precursors of the bony vertebrae. Blocks of sclerotome tissue surround the notochord and the neural tube that extends beneath the notochord. Sclerotome cells merging medially around the notochord become the precursors for vertebral bodies, while dorsal sclerotome cells merge around the neural tube to form precursor neural arches and transverse processes. As the developing spinal nerves begin to grow outward from the developing neural tube, the sclerotome tissue blocks resegment, splitting into condensed caudal and less condensed cranial portions with each portion uniting with adjacent cranial and caudal portions to form the final tissue blocks. The ventral–medial portions of the sclerotome surrounding the notochord are genetically directed to form the centra of the vertebral bodies. Dorsal–lateral portions surrounding the neural tube are directed by a different set of genetic signals to form the transverse processes and neural arches over the spinal cord developing from the neural tube. Resegmentation creates a fissure between the developing vertebral segments, marking the boundaries between them (Fig. B-1.0). The annulus fibrosis forming the intervening vertebral disks develops from separated caudal sclerotome cells, while incorporating residual notochord tissue for the nucleus pulposus as the notochord itself withdraws from the developing vertebral bodies and degenerates. As the precursor intervertebral disks form, similar separated caudal sclerotome cells cluster within the dorsal region of the developing neural arches to give rise to the intervertebral ligaments. Genetic signals direct sequential developmental differentiation of the vertebral segments, while other sets of genetic signals direct the development of changing borders between different portions of the vertebral column.

The seven cervical vertebrae are derived from eight original sclerotomes. As the sclerotomes resegment, splitting into cranial and caudal halves, the cranial half of the first cervical segment joins the adjacent occipital sclerotome of the developing skull base and also provides tissue for the apical tip of the dens developing within the region of the foramen magnum (Fig. A-5.3.0). The caudal half is incorporated into the atlas as it joins the cranial half of the second sclerotome, and also provides tissue for the base of the dens. The eighth cervical sclerotome resegments with the cranial half

Atlas of Developmental Field Anomalies of the Human Skeleton: A Paleopathology Perspective, First Edition. Ethne Barnes.
© 2012 Wiley-Blackwell. Published 2012 by John Wiley & Sons, Inc.

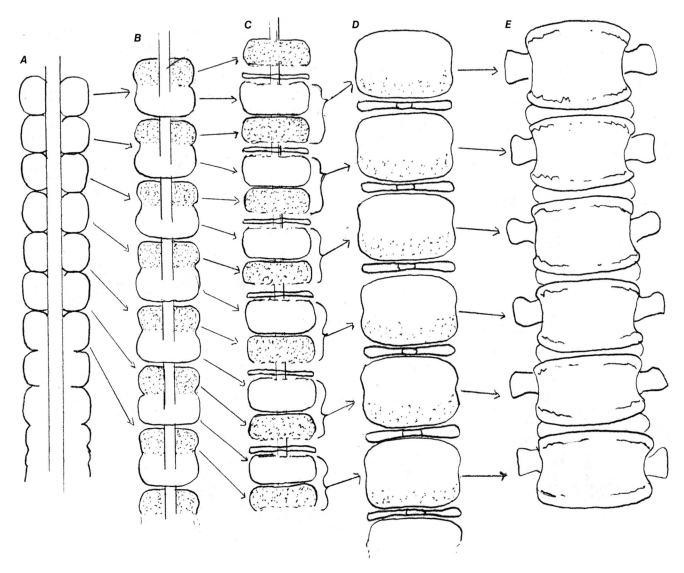

FIGURE B-1.0. Vertebral column development: (*A*) paired somites segregating from the paraxial mesoderm along the sides of the notochord; (*B*) sclerotomes developing from the somites around the notochord with defined cranial and caudal halves; (*C*) resegmentation as caudal and cranial halves of each sclerotome separate and fuse with adjacent caudal and cranial halves as primordial intervertbral disk tissue develops between each newly formed segment; (*D*) final segments with developing intervertbral disks incorporating centralized residual notochord tissue for nucleus pulposus as notochord disappears; (*E*) mature vertebrae.

contributing to the seventh cervical vertebra and the caudal half contributing to the first thoracic vertebra. This sequence of resegmentation events continues down the developing primordial vertebral column, ending with the last caudal segments incorporated into the coccyx.

At birth, the two bony dorsal neural arch segments and anterior body or centrum remain separated but connected by hyaline cartilage programmed during morphogenesis to ossify and unite the separated vertebral parts during growth. The neural arches fuse together in ascending order from the first to third years, followed with fusion to the vertebral bodies in descending order

between the third and eighth years. Secondary ossification of the tips of the vertebral processes and the epiphyseal plates, and union of the sacral segments progress throughout puberty, with final fusions occurring in early adulthood. The atlas vertebral anterior arch remains cartilaginous at birth, ossifying by the end of the first year and uniting with the lateral processes during childhood. The bony axis vertebral body takes shape within the hyaline cartilage model, as mirror lateral ossifications with an upward thrust for the body of the dens come together with the base that ossifies separately to unite during the sixth fetal month. As the lateral parts come together, they form a truncated conical superior tip that generally fills in with bone before birth. A separate ossification center appears in the dens tip during the second year, uniting with the dens body before puberty.

VERTEBRAL COLUMN ANOMALIES

B-1. Vertebral Border Shifting

As discussed earlier under anomalies of the skull base (see Section A-5.3), border shifting occurs at the occipital–cervical border as well as along the vertebral column to the sacral–caudal (S-C) border. Specific structural characteristics that mark both sides of each border between changing vertebral sections are regulated by a combination of genetic codes. Mutant genes can alter the dividing border and change the character of the affected vertebral segment, depending on which way the border is moved. It is important to remember that it is the border that moves, not the vertebral segment. The border moves downward as a caudal shift when it is altered by a "loss-of-function" mutant gene. The border moves upward as a cranial shift when it is altered by a "gain-of-function" mutant gene. Some knock-out mutant genes following the retinoic acid gradient down the developing vertebral column can simultaneously create a caudal shift at one border and a cranial shift at another border (Larsen 2001:106).

Cranial border shifts affect the vertebral segment above the designated border as it takes on the characteristics of the vertebral section that it has joined below it, while caudal shifts affect the vertebral segment below the designated border as it takes on the characteristics of the vertebral section that it has joined above it. Border shifts can be unilateral or bilateral (often asymmetrical), with a variation in expressions ranging from incomplete to complete changing characteristics.

B-1.1. Cranial Shifts of the Cervical–Thoracic (C-T) Border

This primarily affects the anterior costal portion of the transverse processes of the seventh cervical vertebra as it is incorporated into the rib-bearing thoracic section of the vertebral column (Fig. B-1.1.1). The costal portions can remain separate from the dorsal transverse portion and develop into ribs or rib-like expressions, depending on the force of the cranial shifting. Complete shifting produces a separate, articulating rib (class IV cervical rib) imitating a first thoracic rib (Fig. B-1.1.2). However, the cervical rib appears more constricted and narrower than the first rib, and the width and length can vary from 20 to 85 mm. They may terminate in a free end, or connect by way of a ligamentous band with the sternal manubrium or first costal cartilage, but most

often are directed downward to articulate or fuse with the first rib. Extra rib or rib-like bony extensions have also been recorded on rare occasions for the fourth, fifth, and sixth cervical vertebrae (Bailey 1974; Schmorl and Junghanns 1971) that also contain costal segments on the transverse processes that can ossify as small separate or attached entities influenced by cranial border shifting. The cervical rib may lack an articulating joint (Fig. B-1.1.3), presenting as a greatly enlarged extension of the transverse process (class III cervical rib). Sometimes cervical rib expression is limited to blunt, bony extensions 40–50 mm long (class II cervical rib), or just a small tubercle or small separate ossicle (Fig. B-1.1.4) at the end of the transverse process (class I cervical rib). The classifications were defined by Gruber nearly 100 years ago (Honeij 1920). Cervical ribs that extend far forward can exert pressure on the brachial plexus, subclavian artery and vein, and or the eighth cervical and first thoracic nerves. Trauma or functional stress can create greater compression, obstruct blood circulation, and produce neuralgic pains, numbness, and muscle weakness that can impair the affected hand and arm muscles with atrophy and motor loss.

B-1.2. Caudal Shifts of the C-T Border

As the border moves downward, the first thoracic vertebra tries to incorporate into the cervical section but is unable to loose its ribs. However, the ribs and transverse processes are reduced, and rudimentary cervical-like transverse foramina are often expressed (Fig. B-1.2). The transverse processes of the seventh cervical vertebra may appear shorter than those above it. The modified first thoracic ribs vary in size, are often quite small (less than 30 mm long) and abnormal in shape (but still wider than the cervical ribs), and frequently articulate or fuse to the second ribs, or connect with the sternal manubrium through a ligamentous band instead of costal cartilage. The second rib costal cartilage moves downward to attach to the sides of the mesosternum instead of at the designated manubrium–sternal interface. The abnormal thoracic rib is more likely to create more severe pressure problems on the brachial plexus and subclavian blood vessels than the cervical rib (White et al. 1945).

B-1.3. Cranial Shifts of the Thoracic–Lumbar (T-L) Border

Complete shifting forces the twelfth thoracic vertebra to join the lumbar section, reducing or eliminating its ribs. Most often, the twelfth ribs are not recoverable in archaeological settings, but small or absent rib facets on the twelfth thoracic vertebra will reflect hypoplasia or agenesis of ribs affected by cranial shifting.

Transitional apophyseal facets on the twelfth thoracic vertebra designate the changing border between

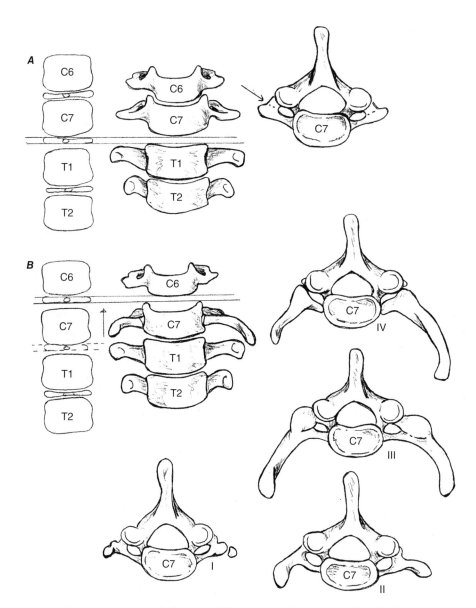

FIGURE B-1.1.1. Cranial shifts at the cervical-thoracic (C-T) border: (*A*) designated border between the seventh cervical and first thoracic vertebrae (note the arrow pointing to the costal process of the seventh cervical); (*B*) cranial border shifting upward between the sixth and seventh cervical vertebrae, class IV complete cervical ribs, class III cervical ribs without articulating joints, class II cervical rib expression—blunt extensions, class I cervical rib expressions—attached and separate bony tubercles.

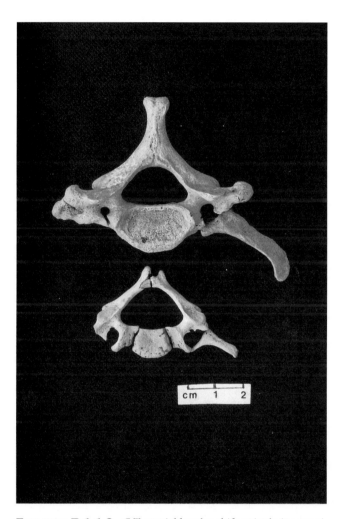

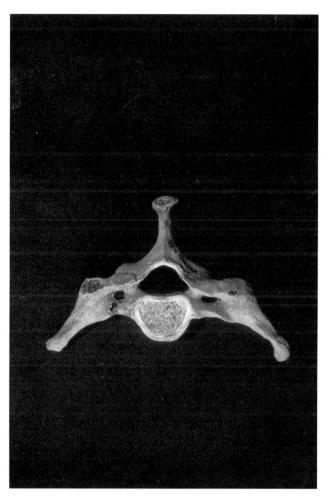

FIGURE B-1.1.2. C-T cranial border shift articulating cervical ribs: adult female and infant (ca. 16 months) unilateral left, Frankish Corinth, Greece.

FIGURE B-1.1.3. C-T cranial border shift jointless cervical ribs: long bony extensions lacking articulating joints, adult (NMNH 271896), Amoxiumqua, NM.

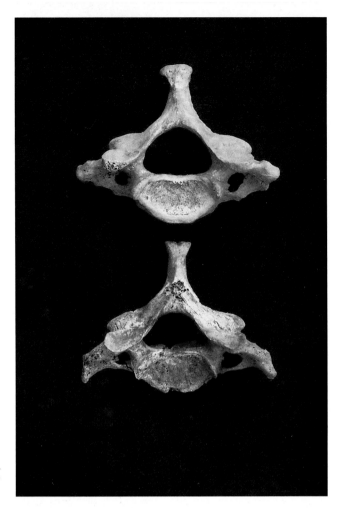

F I G U R E B-1.1.4. C-T cranial border shift mild expression cervical ribs: blunt extensions (bottom) compared with normal C7 (top), adolescent female, Frankish Corinth.

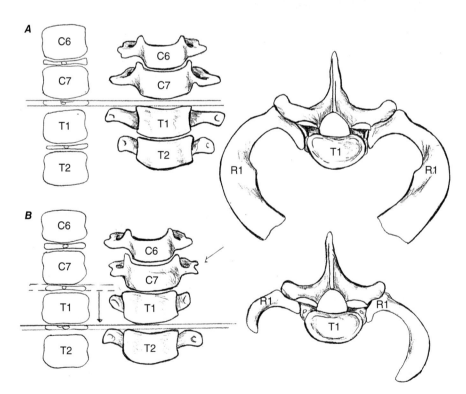

F I G U R E B-1.2. Caudal shifts at the cervical–thoracic (C-T) border: (*A*) designated border between the seventh cervical and first thoracic vertebrae; (*B*) caudal border shifting downward reducing transverse processes of the first thoracic and first ribs, with rudimentary transverse foramina (arrow points to reduced transverse processes on the seventh cervical).

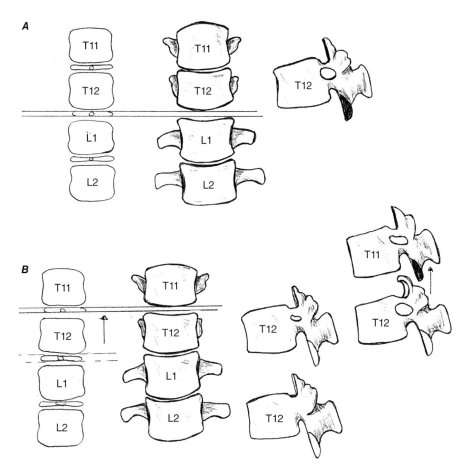

F IGURE B-1.3. Cranial shifts at the thoracic–lumbar (T-L) border: (*A*) designated border between the twelfth thoracic and first lumbar vertebrae (T12 transitional facets darkened); (*B*) cranial border shifting upward between the eleventh and twelfth thoracic vertebrae with reduced rib facet for the twelfth rib hypoplasia, absent rib facet for the twelfth rib agenesis, mild shifting transitional facets moving upward to the eleventh thoracic (darkened facets).

the thoracic and lumbar sections of the spine: curved inferior lumbar-like facets with superior flat thoracic facets. Mild cranial border shifting moves the transitional facets up to the eleventh thoracic vertebra, placing the curved lumbar-like facets interfacing between the eleventh and twelfth thoracics (Fig. B-1.3).

B-1.4. Caudal Shifts of the T-L Border

The first lumbar vertebra takes on the characteristics of the thoracic section as it is placed above the designated border by downward border shifting (Fig. B-1.4.1). The transverse processes transform into articulating ribs—lumbar ribs, or elongated rib-like extensions that fail to detach. Lumbar rib expressions can vary in shape and size, from small, articulating costal tubercles to complete articulating ribs up to 70 mm long (Fig. B-1.4.2). Lumbar ribs differ from rudimentary twelfth ribs in that they are usually thick and blunt tipped and horizontally directed compared with the more slender, tapered-tipped and obliquely angled upward twelfth ribs. Larger

lumbar ribs can cause back pain or soreness, especially with functional stress or trauma.

Mild caudal border shifting can move the transitional facets from the twelfth thoracic vertebra down to the first lumbar vertebra. The curved interfacing apophyseal facets between the twelfth thoracic and first lumbar are replaced with flat thoracic-like facets, leaving the first lumbar with flat superior facets and inferior curved facets, and the twelfth thoracic with both superior and inferior flat facets.

B-1.5. Cranial Shifts of the Lumbar–Sacral (L-S) Border

Complete upward shifting of the border forces the last lumbar vertebra to become part of the sacrum (sacralized lumbar). Assimilation into the sacrum can vary in expression when incomplete. The pedicles and transverse processes may transform into wide, ala-like or bifid processes that may or may not articulate with the sacral alae, and they can articulate with the ilium if

FIGURE B-1.4.1. Caudal shifts at the thoracic–lumbar (T-L) border: (*A*) designated border between the twelfth thoracic and first lumbar vertebrae; (*B*) caudal border shifting downward creates the first lumbar rib facets with blunt-tipped lumbar ribs, mild shifting transitional facets (darkened) move downward from the twelfth thoracic to the first lumbar.

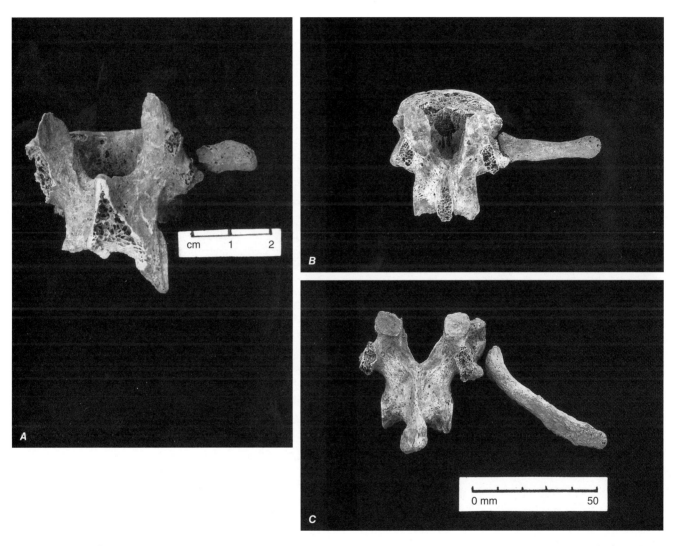

FIGURE B-1.4.2. T-L caudal border shift lumbar ribs: (*A*) adult female, Pietri, Corinth; (*B*) adult female, Frankish Corinth; (*C*) adult male, Pietri, Corinth, Greece.

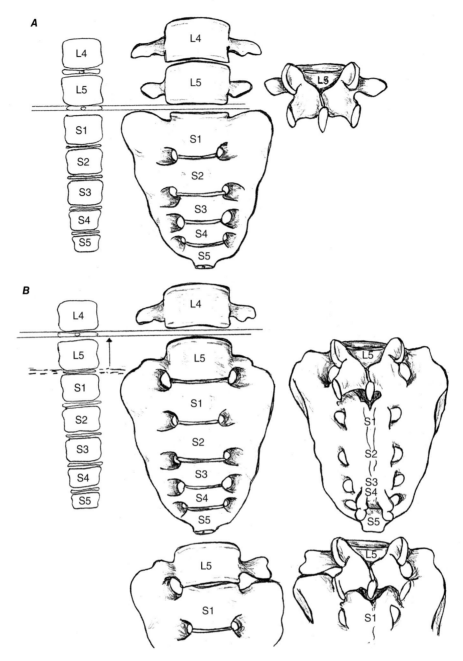

FIGURE B-1.5. Cranial shifts at the lumbar–sacral (L-S) border: (*A*) designated border between the fifth lumbar vertebra and the first sacral segment; (*B*) cranial border shifting upward between the fourth and fifth lumbar vertebrae, complete and partial sacralization of the fifth lumbar.

long and directed laterally. Often, the sacralized lumbar retains some form of its dorsal apophyseal articulations with the sacrum. Incomplete anterior fusion can appear as a marginal cleft between the bodies of the lumbar and first sacral segments. Incomplete unilateral expressions of sacralized lumbar vertebra can appear asymmetrical and more prone to produce low back pain and sciatica, and may lead to curvature and rotation of the lower spine (Fig. B-1.5).

B-1.6. Caudal Shifts of the L-S Border

This causes the first sacral segment to take on lumbar characteristics, and with complete shifting, the sacral segment separates completely from the sacral body (lumbarized sacral), with its alae somewhat transformed into transverse processes that may or may not articulate with the remaining sacral alae. The lumbarized sacral body is short and wide, and articulating apophyseal processes are usually rudimentary. Incomplete shifts lead to partial separation. Sometimes, only rudimentary dorsal apophyseal articulations or anterior clefting between the first and second sacral segments appears with mild shifting. With complete lumbarization, the alae of the second sacral segment of the remaining sacrum generally appear higher than the level of the sacral body. Incomplete unilateral expressions of lumbarized sacral segment can cause similar back problems as with the asymmetrical sacralized lumbar (Fig. B-1.6).

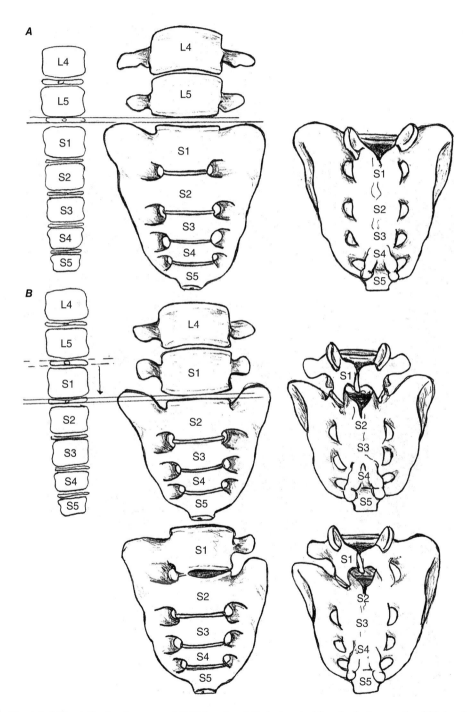

FIGURE B-1.6. Caudal shifts at the lumbar–sacral (L-S) border: (*A*) designated border between the fifth lumbar vertebra and the first sacral segment; (*B*) caudal border shifting downward between the first and second sacral segments, complete and partial lumbarization of the first sacral segment.

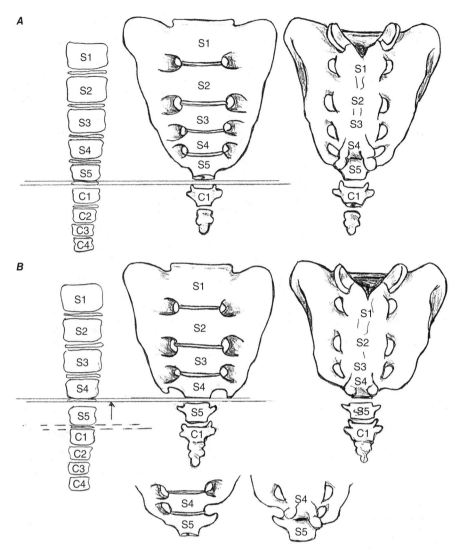

FIGURE B-1.7. Cranial shifts at the sacral–caudal (S-C) border: (*A*) designated border between the fifth sacral segment and the first caudal segment of the coccyx; (*B*) cranial border shifting upward between the fourth and fifth sacral segments, complete and partial caudalization of the fifth sacral segment.

B-1.7. Cranial Shifts of the Sacral–Caudal (S-C) Border

As the border between the last sacral segment and first caudal segment of the coccyx moves upward, the last sacral segment separates completely or partially to join the coccyx (caudalized sacral), leaving the lateral sacral foramina between the last two sacral segments absent or incomplete. Partial or incomplete separation is more common than complete separation (Fig. B-1.7).

B-1.8. Caudal Shifts of the S-C Border

The first caudal segment of the coccyx is forced to join the sacrum (sacralized caudal) completely or partially

as the border shifts downward. Complete assimilation produces extra lateral foramina in the sacrum, while the caudal cornua generally remain separate from the sacral dorsal protrusions. Sometimes, only the caudal body unites with the sacrum (Fig. B-1.8).

B-2. Extra Vertebral Segment (Transitional Vertebra)

Sometimes an extra vertebral segment formed from an extra pair of somites is added to the vertebral column. The extra vertebral segment becomes a transitional vertebra appearing between borders of the lower spine:

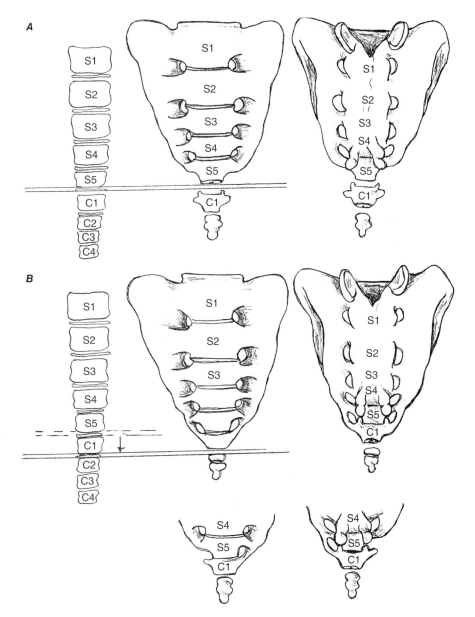

FIGURE B-1.8. Caudal shifts at the sacral–caudal (S-C) border: (*A*) designated border between the fifth sacral segment and the first caudal segment of the coccyx; (*B*) caudal border shifting downward between the first and second caudal segments of the coccyx, complete and partial sacralization of the first caudal segment.

the T-L, L-S, or S-C borders (Fig. B-2). An extra vertebral segment appearing at the C-T border of the upper spine is extremely rare. The extra vertebral segment adds confusion to deciphering border shifting as it can transition to either side of the border, taking on the characteristics of either side. Careful counting of all vertebral segments is necessary to determine the presence of an extra vertebral segment.

B-3. Cleft Neural Arch

This occurs with failure of osseous fusion of the two sides of the neural arch. Underlying developmental delays result in hypoplasia or aplasia of both (bilateral) or one (unilateral) of the bony neural arch halves. Failure of joining ossification leaves a bony cleft that can vary in size as fibrocartilage holds the two halves

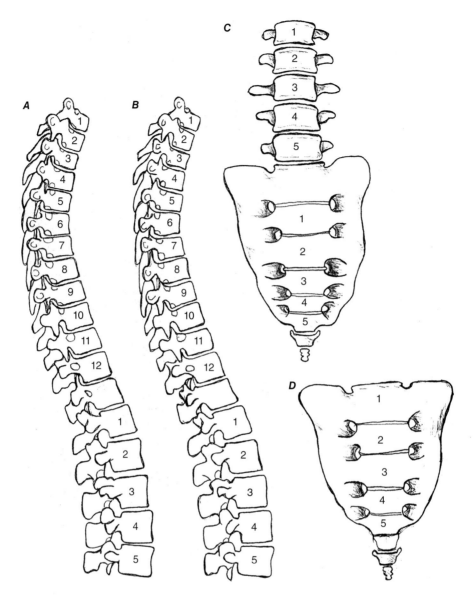

F I G U R E B-2. Extra vertebral segment (transitional vertebra): (*A*) extra vertebral segment at the thoracic–lumbar border expressed as the thoracic vertebra (note rib facet); (*B*) extra vertebral segment at the thoracic–lumbar border expressed as the lumbar vertebra (no rib facet); (*C*) extra vertebral segment at the lumbar–sacral border expressed as the sacral segment; (*D*) extra vertebral segment at the sacral–coccyx border expressed as the caudal segment.

together. The associated inferior apophyseal facet is also altered in size. Cleft variations range from slight bifurcation of the laminae and spinous process to cleft laminae with or without the spinous process (Fig. B-3.1). Very rarely, aplasia of one-half of the neural arch results in an asymmetrical wide cleft (Fig. B-3.2), while even rarer absence of the whole bony neural arch can occur when both sides suffer aplasia. Unilateral hypoplasia affecting only one side can leave the other side developing a spinous process that often overlaps the smaller half. Mild to moderate forms of cleft neural arch are not uncommon and are nonpathological, unlike the neural tube defect spina bifida that interferes

with neural arch development (see Section B-6). Cleft neural arch has often been misidentified as spina bifida occulta, the lesser neural tube defect in spinal cord development. The affected neural arches in cleft neural arch, unlike the defect with spina bifida, follow the directed pathway set in the fibrocartilage connection. Defective neural arches with spina bifida lack the fibrocartilage connecting pathway (see Section B-6). Cleft neural arch as a separate entity from neural tube defect was first recognized by Theodore A. Willis (1929).

Clefting can affect the neural arches of any vertebra but rarely involves more than one or two presacral vertebrae in the same individual. The first sacral segment

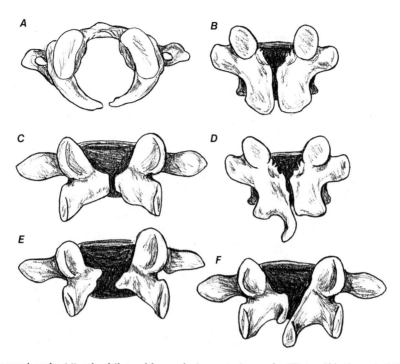

FIGURE B-3.1. Cleft neural arch: (*A*) atlas bilateral hypoplasia posterior arch; (*B*) twelfth thoracic bilateral hypoplasia without the spinous process; (*C*) fifth lumbar bilateral hypoplasia without the spinous process; (*D*) twelfth thoracic unilateral hypoplasia with the spinous process; (*E*) fifth lumbar unilateral hypoplasia without the spinous process; (*F*) fifth lumbar unilateral hypoplasia with the spinous process.

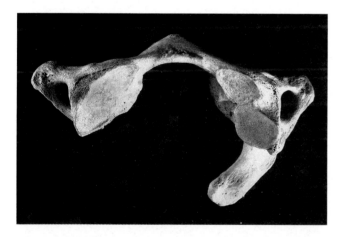

FIGURE B-3.2. Cleft neural arch—atlas: unilateral aplasia with a wide asymmetrical cleft, Alberta, Canada (contributed by C.F. Merbs).

is often affected by clefting that may extend into the second sacral segment. Complete clefting or bifurcation of the sacrum is not unusual, extending into the designated sacral hiatus that often extends upward through the fourth segment (Figs. B-3.3 and B-3.4).

The next most commonly affected vertebral segments are the posterior arch of the cervical atlas (Fig. B-3.5) and neural arch of the fifth lumbar. When the atlas posterior arch is affected, the posterior bony tubercle (rudimentary spinous process) is usually absent. Clefting

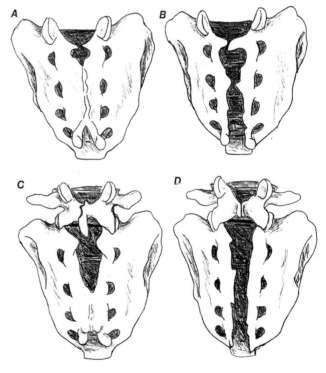

FIGURE B-3.3. Cleft sacral neural arches: (*A*) cleft S1; (*B*) complete irregular cleft; (*C*) cleft S1 and S2 with cleft L5; (*D*) complete cleft with slight cleft L5.

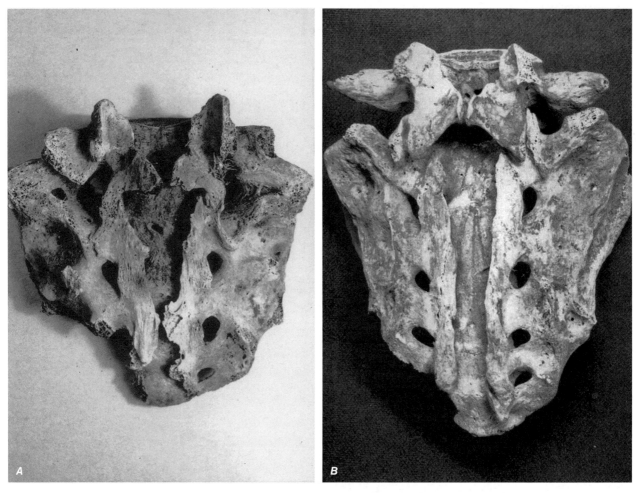

FIGURE B-3.4. Cleft sacral neural arches—complete: (*A*) adult male (NMNH 228941), Otowi, NM; (*B*) with cleft L5, adult female (NMNH 263005), Puye, NM.

of the thoracic vertebrae is rare, affecting primarily the lower thoracics. Clefting of the other cervicals (Figs. B-3.6) and upper thoracics is unusual. Atypical clefting of a portion of an arch is very unusual (B-3.7). Clefting caused by neural arch unilateral hypoplasia, especially in the lumbar area, often leads to a unilateral spondylolysis (Fig. B-3.8) caused by uneven load bearing stress on the affected vertebral segment.

B-4. Cleft Atlas Anterior Arch

This generally occurs with morphogenetic programmed failure of the anterior arch to ossify from its cartilage precursor by the end of the first year. In order to maintain stability when ossification of the anterior arch fails, the two lateral masses usually form bony extensions that may or may not meet midline and fuse. Sometimes, bifurcation of the anterior arch develops when two

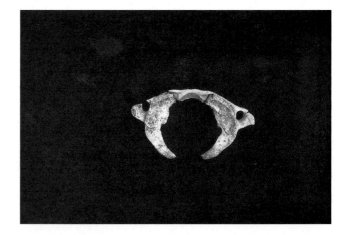

FIGURE B-3.5. Cleft neural arch—atlas: bilateral hypoplasia, child 5-6 years, Frankish Corinth, Greece.

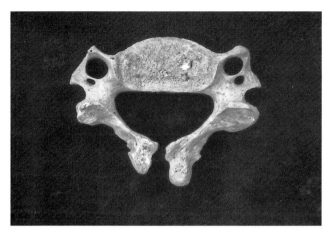

FIGURE B-3.6. Cleft neural arch—C6: unilateral left hypoplasia, child, 11-12 years, Frankish Corinth, Greece.

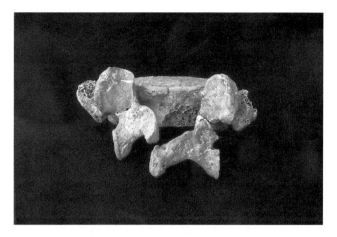

FIGURE B-3.8. Cleft neural arch—lumbarized S1: unilateral left hypoplasia with unilateral right spondylolysis, adult male, Byzantine Petras, Crete.

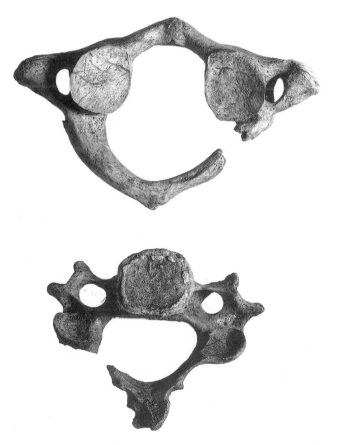

FIGURE B-3.7. Cleft neural arch atypical forms: atlas and C4, adults, Santa Rosa Island, CA (contributed by C.F. Merbs).

separate ossification centers develop within the cartilage precursor and fail to coalesce. One of these two ossification centers may fail to appear, resulting in unilateral aplasia, with only an incomplete bony anterior arch forming. Cleft anterior arch often appears with a cleft posterior arch (Fig. B-4).

B-5.1. Notochord Defect: Sagittal Cleft Vertebra

This results from the failure of a portion of the notochord to recede and disappear from the developing embryonic vertebral segment, disrupting the formation of the centrum in a variety of expressions, depending on the location of the renegade tissue (Fig. B-5.1.1). The retained notochord tissue turns fibrous, persisting as a linear tissue streak, tissue pocket (Fig. B-5.1.2), or as a major sagittal tissue defect dividing the centrum, forming a sagittal cleft (Figs. B-5.1.3 and B-5.1.4). The neural arch is unaffected.

Major notochord tissue disturbances divide the developing centrum along the sagittal plane into two laterally wedge-shaped halves (Fig. B-5.1.5), often asymmetrical in form with decreased or absent anterior portions (referred to as "butterfly vertebra" on radiographs). The gaps filled with fibrous tissue from the notochord often inhibit the development of the centrum halves with hypoplasia of one or both. Sometimes, bony strands or a bony bridge reach across the cleft to connect the two halves. Adjacent vertebral bodies compensate for the depressed gaps within the affected vertebral body with bony buildup on their interfacing surfaces.

Both linear streaks and tissue pockets of retained notochord tissue appear indented on the surface of the affected end plate. The retained notochord tissue defect can affect either the superior or inferior end plate or both, or penetrate through the vertebral body from one end plate to the other. The linear streak generally appears along or near the midsection of the vertebral body, but can angle off from the sagittal plane. It is usually shallow and narrow but sometimes very wide, and may or may not penetrate through the posterior and/or anterior margins of the vertebral body. Pockets of retained notochord tissue can deviate away from the

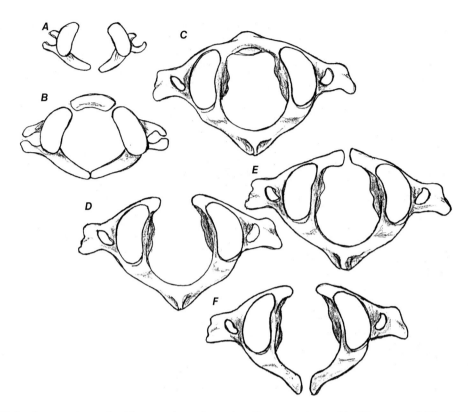

FIGURE B-4. Cleft atlas anterior arch: (*A*) normal newborn (cartilage anterior arch); (*B*) young child with bony anterior arch not yet fused; adult—(*C*) adult atlas; (*D*) aplasia bony anterior arch; (*E*) aplasia bony anterior arch, replaced with bony extensions from lateral processes; (*F*) cleft anterior and posterior arches.

midsection of the vertebral body. These pockets appear large or small, somewhat rounded, as shallow indentations or complete apertures penetrating through the vertebral body (Anderson 2003; Merbs 2004).

Coronal cleft vertebra, a transverse cleft appears when entrapped notochord tissue within the precursor cartilaginous centrum interferes with merging of the dorsal and anterior portions in the cartilaginous stage prior to ossification. This cleft can be expressed as irregular transverse or ovoid depressions, usually located in the lumbar spine or sometimes in the thoracic spine but never in the cervical spine. This is very rare in adults, but often seen in infancy, disappearing with the growth of affected vertebral bodies as the infantile fibrous cleft ossifies (Warkany 1971).

B-5.2. Notochord Defect Diastematomyelia

This is an extremely rare disturbance involving aberrant notochordal cells straying from their programmed migratory pathway during morphogenesis, usually within the thoracolumbar spine. The renegade cells become trapped in neuroectodermal tissue and form a pedicle within the developing spinal cord that connects with the notochord surrounded by the developing centrum. The inductive nature of the renegade notochord cells attracts undifferentiated mesenchymal cells that in turn develop into precartilaginous cells as the pedicle attaches to the posterior surface of the centrum with the other end extending to the developing neural arch. The pedicle bifurcates the developing spinal cord tissue and its covering membranes, and as development proceeds, the pedicle either evolves into fibrocartilage or ossifies into a bony spicule (Fig. B-5.2). The bifurcated spinal cord expands the spinal canal, and this forces the pedicles of the affected vertebra to spread. Motor and sensory disturbances of variable severity begin to appear between 2 and 10 years of age as growth and development place drag on the spinal cord caused by the defect (Harris 1959).

B-6. Neural Tube Defect Spina Bifida

This is caused by the disruption of the developing spinal cord (spinal dysraphism) occuring during or shortly after closure of the posterior neuropore primarily in the L-S region (Lemire 1988). Disturbance in neural tube development is linked to epigenetic folate metabolic dysfunction (Yates et al. 1987). Maternal zinc and selenium deficiencies also appear to affect neural tube morphogenesis (Myrianthopoulos and Melnick 1989; Zimmerman and Lozzia 1989).

Failure of the posterior neuropore (neurulation defect) to close properly displaces a portion of the

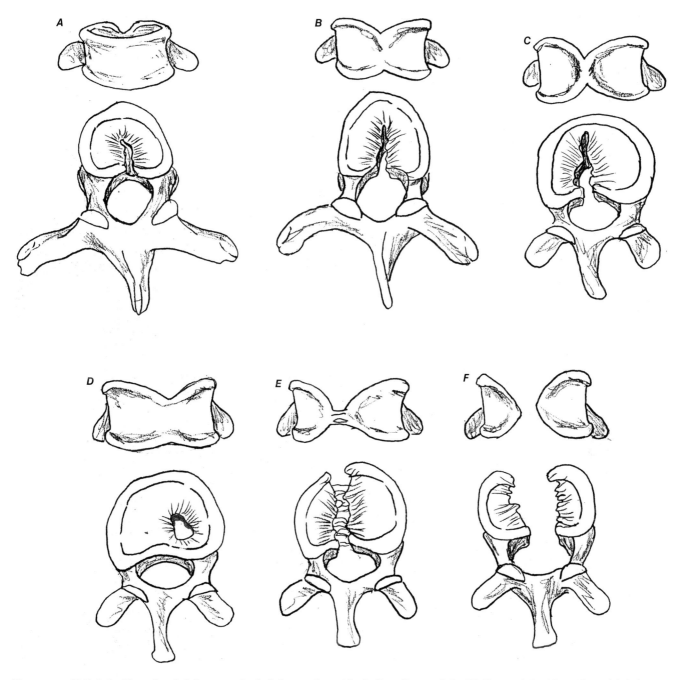

FIGURE B-5.1.1. Notochord defect—sagittal cleft vertebra: (*A*) shallow linear cleft; (*B*) linear cleft; (*C*) unilateral left linear cleft; (*D*) cleft pocket; (*E*) complete cleft with connecting bony spicules; (*F*) complete cleft.

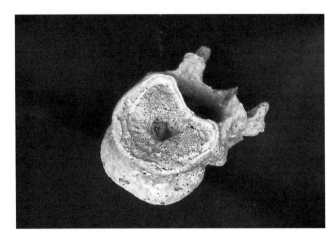

FIGURE B-5.1.2. Notochord defect—sagittal cleft pocket: T12, adult female, Byzantine Petras, Crete.

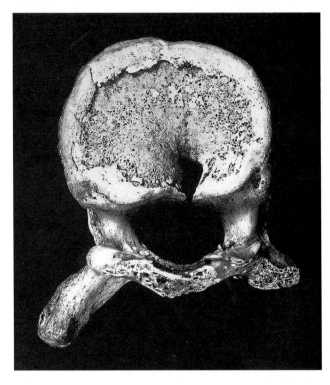

FIGURE B-5.1.4. Notochord defect—sagittal cleft vertebra: T10, adult, Pastolik, Alaska (contributed by C.F. Merbs).

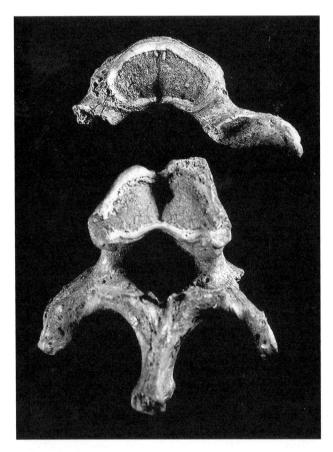

FIGURE B-5.1.3. Notochord defect—sagittal cleft vertebrae: T1 and T3, adult, Kuaua, NM (contributed by C.F. Merbs).

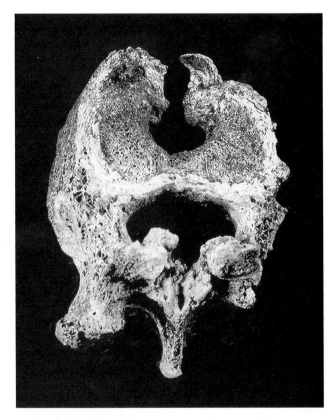

FIGURE B-5.1.5. Notochord defect—sagittal cleft vertebra: T12, adult, Chavez Pass, AZ (contributed by C.F. Merbs).

developing spinal tissue and nerve roots outside the vertebral canal (myelodysplasia), forming a cyst covered by membrane known as a meningomyelocele (myleo-meningocele). The emerging cyst disrupts the development and fusion of the precursor neural arches of the adjacent vertebrae. The pedicles grow thin and stunted, while the laminae appear deformed or absent without development of a spinous process, and the apophyseal

facets may be distorted or absent. Developing neural arches above or below the disturbance can also be affected to some degree. The bulging spinal tissue defect raises the affected neural arch halves to accommodate the developing soft tissue intrusion known as spina bifida cystica (Epstein 1976). Various lower body neurological disabilities, especially paraplegia, are always present with meningomyelocele, and hydrocephalus often develops when the defect interferes with the flow of the cerebrospinal fluid (CSF) throughout the central nervous system, increasing the volume and pressure on the cerebral ventricles. It is highly unlikely that this type of neural tube defect (NTD) would survive the neonatal period in paleopathology.

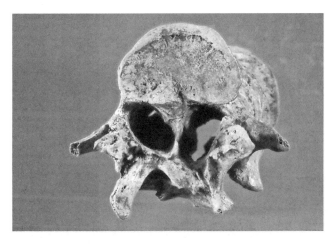

FIGURE B-5.2. Notochord defect—diastematomyelia: bony spicule dividing the spinal canal, fifth lumbar, young adult, Negev Desert, Israel (Paleopathology Club contribution from Dr. Baruch Arensburg).

Neural tube defects occurring after the closure of the posterior neuropore (postneurulation defect) have a better chance of surviving into adulthood with less severe pathology. The developing spinal tissue and nerve roots remain within the vertebral canal, but the membranous covering of the spinal cord—the meninges (external dura mater and middle arachnoid)—protrude outward, disrupting adjacent developing neural arches. This leads to the formation of a fluid-filled spina bifida cystica that fuses with the overlying ectoderm forming a skin-covered meningocele that is often smaller than the meningomyelocele. Again, the most common area affected is the L-S spine with similar neural arch disturbances as the more severe NTD (Figs. B-6.1.1 and

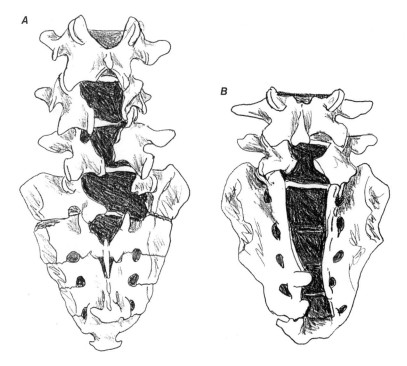

FIGURE B-6.1.1. Neural tube defect spina bifida: (*A*) disrupting neural arch development of the fourth and fifth lumbars and the first sacral segment of a prehistoric young adolescent from Florida (drawn from Dickel and Doran 1989); (*B*) disrupting neural arch development of the fifth lumbar and sacrum of a young adult female from the Federal Pathologic Anatomy Museum in Vienna, Austria (drawn from Ortner 2003:465).

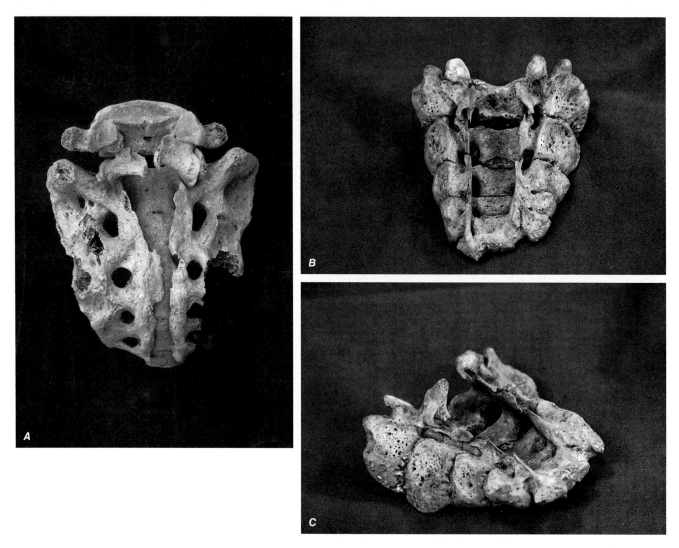

FIGURE B-6.1.2. Neural tube defect spina bifida versus cleft neural arch: (*A*) cleft sacral neural arches and fifth lumbar, adult male, Frankish Corinth, Greece; (*B,C*) neural tube defect sacral spina bifida, adolescent male (NMNH 345338), Alaska (lumbar spine not available except for a piece of the fifth lumbar with signs of disrupted neural arch).

B-6.1.2B,C). When several vertebrae are affected, the defect may appear fusiform—tapering at both ends with the widest part in the middle. Pedunculated meningocele is less disruptive with corresponding smaller neural arch defect. Depending on the location and severity of the NTD, neurological symptoms can range from moderate to severe, with minor defects shielded beneath a layer of fat below the skin surface, often defined outwardly by a dimple or tuft of hair—spina bifida occulta. This generally occurs in the midsacral or L-S regions. Adjacent neural arches reflect the underlying disturbance similar to other NTDs (Ruge and Wiltse 1977). The major differences between cleft neural arch from the failure of osseous fusion and clefts formed by NTD spina bifida: Cleft neural arches remain within their designated alignment with normal vertebral canal space,

often involving only one, sometimes two presacral vertebrae; NTD spina bifida arches are thin, pushed outward with widened vertebral canal and distorted pedicles, usually involving more than two presacral vertebrae. Both can produce a completely cleft sacrum but differ in alignment of the cleft arches (Figs. B-6.1.2 and B-6.2), wide and pushed outward with NTD, flat, often with irregular borders with cleft neural arch (see Section B-3).

B-7. Hemivertebra: Hemimetameric Shifts

Developing somite pairs—hemimetameres—meet midline and fuse to form the sclerotomes in a descending cranial–caudal direction. If one of the paired somites is

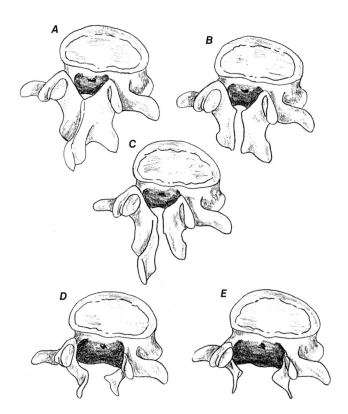

FIGURE B-6.2. Neural arch cleft versus spina bifida cleft: (A) normal; (B,C) cleft neural arch; (D,E) neural tube defect spina bifida cleft (note the difference in width of the spinal canal).

slow to move to the midline at the critical threshold moment for fusion with its partner, it will shift downward one segment, trying to pair up with the next available hemimetamere approaching its threshold time for fusion, creating disorder in the developing vertebral segments (Fig. B-7.1). The out-of-sink hemimetameric segment fails to converge with a matching segment and develops into a lateral wedge-shaped hemivertebra, with a normal halved neural arch that does not cross the midsagittal plane of the vertebral column (Tsou et al. 1980). Thoracic hemivertebrae usually have normal ribs attached to them.

Contralateral double shift hemivertebrae, the most frequently occurring hemimetameric shift defect, involves double shifting of two opposing pairs of somites. The hemimetameric double shifts are usually separated by one or more normal developing vertebral segments, double balanced to offset any spinal deformity (Figs. B-7.1A and B-7.2). Bilateral hemivertebrae develop when two opposing hemimetameres shift downward at the same time but fail to meet midline on time to fuse together (Fig. B-7.1B), sometimes fusing only partially.

Solitary hemivertebra usually occurs when an extra somite hemimetamere appears, or one-half of a vertebral

segment fails to develop without a matching partner (Figs. B-7.1C and B-7.3). Unilateral double hemivertebrae appear when two hemimetamere shifts take place on the same side (Fig. B-7.1D). Both solitary and unilateral double hemivertebrae create unbalanced development of the vertebral column, leading to scoliosis. Adjacent developing vertebral segments often compensate for hemimetamere shifting with corresponding deformities and often fuse to the defective segments.

B-8. Lateral Hypoplasia/Aplasia

One of the paired somite hemimetameres either fails to reach normal size, or fails to develop without triggering hemimetamere shifting. The smaller hemimetamere meets its full-sized partner midline on time and fuses as programmed, forming a laterally wedged vertebral segment (Fig. B-8.1.1) varying from slight (Figs. B-8.1.2) to a more severe misshapen form (Fig. B-8.1.3). Very rarely, one side of a hemimetamere pair fails to develop, leaving the other side to develop as a lateral wedge-shaped segment that crosses the midline, unlike a hemivertebra that does not cross the midline.

Multiple vertebral segments are often affected with varying degrees of scoliosis. Lateral hypoplasia affecting more than one vertebral segment can appear all on the same side of the vertebral column with curvature of the spine (Fig. B-8.1.4), or they can develop on both sides of the vertebral column—contralateral hypoplasia—forming an S-shaped curvature of the spine (Figs. B-8.2.1 and B-8.2.2).

When multiple unilateral hypoplastic hemimetameres develop, the affected apophyseal joints are missing and the laminae frequently coalesce into a bony mass to form a bony protusion known as a postlateral bar (Fig. B-8.1.5) (Epstein 1976; Keim and Hensinger 1989). Developing ribs associated with lateral hypoplasia/aplasia fuse together near their vertebral junctions with or without the formation of costovertebral joints with the thoracic rib cage usually deformed.

B-9. Ventral Hypoplasia/Aplasia

This appears with underdevelopment or absence of the ventral precursor cartilaginous center of the centrum with the dorsal cartilaginous center unaffected as the embryonic vertebral body takes form. Normally, as soon as the two cartilaginous centers appear, they quickly coalesce into one to become the forerunner of the future ossification center for the vertebral body. Hypoplasia of the ventral cartilaginous center leads to a ventrally oriented wedge-shaped vertebral body as the dorsal part develops as programmed (Figs. B-9.1–B-9.3). Rarely, aplasia of the ventral embryonic cartilage leaves the ventral portion of the vertebral body missing with

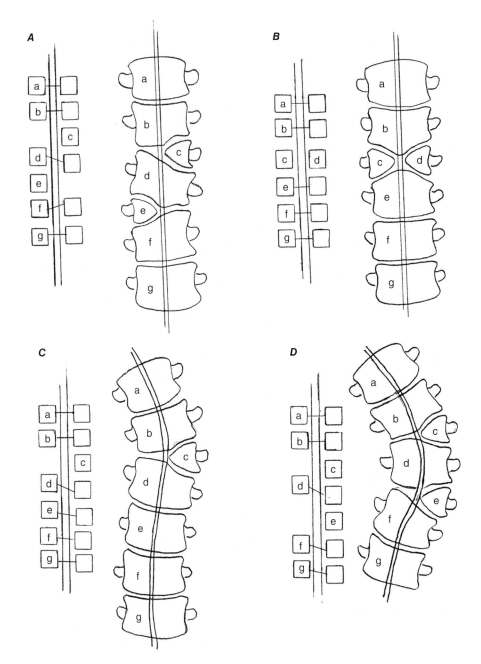

FIGURE B-7.1. Hemivertebra—hemimetameric shifts: (*A*) contralateral double shift; (*B*) bilateral shift; (*C*) solitary hemivertebra; (*D*) unilateral double shift.

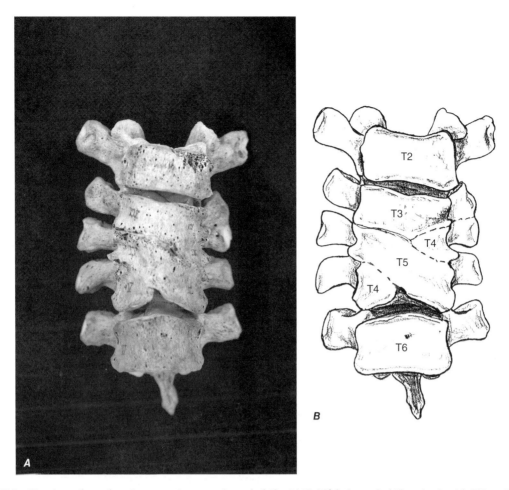

FIGURE B-7.2. Hemivertebra—hemimetameric contralateral shift: (*A,B*) T4 balanced shift united with T5 and T3, forming a multiple block vertebra, adult female (NMNH 262939), Puye, NM.

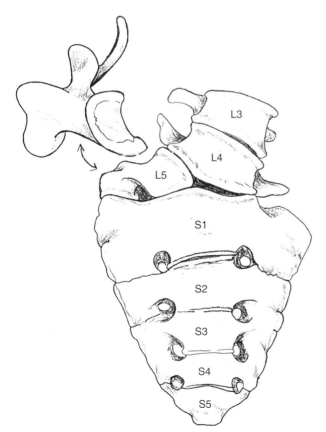

FIGURE B-7.3. Hemivertebra—solitary hemimetamere: L5 half with agenesis of the other half, young adult female (NMNH 381243), Quarai, NM.

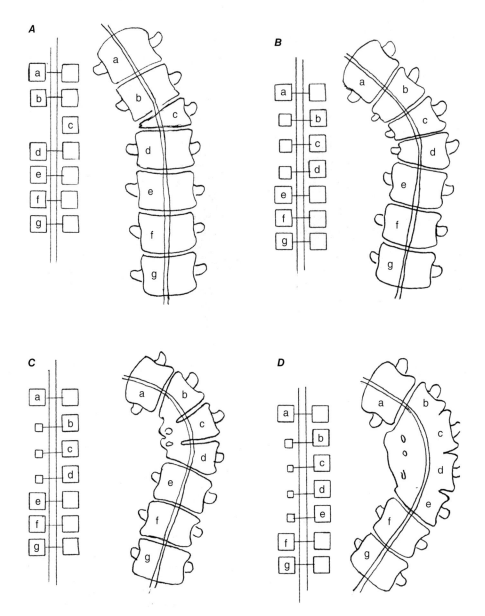

Figure B-8.1.1. Lateral hypoplasia/aplasia: (*A*) solitary unilateral aplasia; (*B*) unilateral multiple hypoplasia; (*C*) unilateral multiple hypoplasia with coalescence into postaleral bar; (*D*) severe unilateral multiple hypoplasia with coalescence into a large postlateral bar.

F I G U R E B-8.1.2. Lateral hypoplasia—mild: right side T10 (T9 to T12 shown), adult male, Frankish Corinth, Greece.

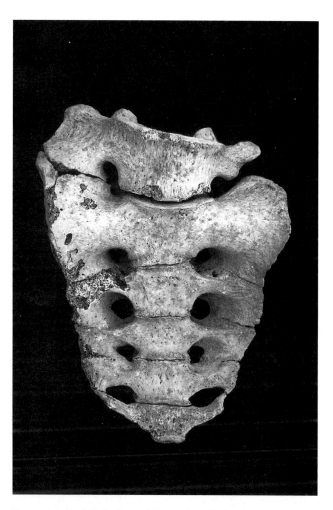

F I G U R E B-8.1.3. Lateral hypoplasia: left side sacralized L5, young adult female, Frankish Corinth, Greece.

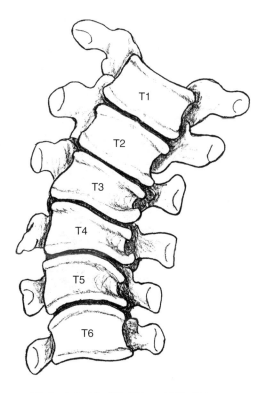

F I G U R E B-8.1.4. Lateral hypoplasia—multiple: left side T3, T4, and T5, Chinese cannery worker (NMNH 364426), Alaska.

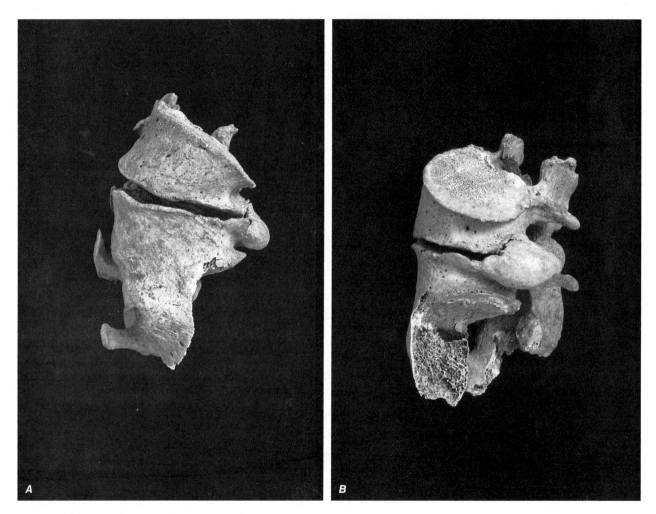

FIGURE B-8.1.5. Lateral hypoplasia with a postlateral bar: (*A*) anterior view and (*B*) left side view—L1 and L2 (all other vertebrae missing; note the bony sheath extending from L2 that covered L3 most likely also affected by left lateral hypoplasia; also note the thin upward right transverse processes from muscle pull), adult, Medieval Well, Corinth, Greece.

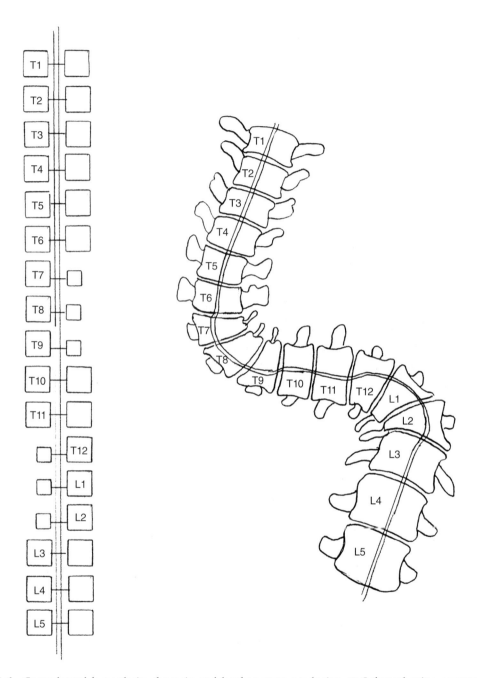

FIGURE B-8.2.1. Contralateral hypoplasia: thoracic and lumbar areas producing an S-shaped spine accompanied by convex rotations of severe scoliosis.

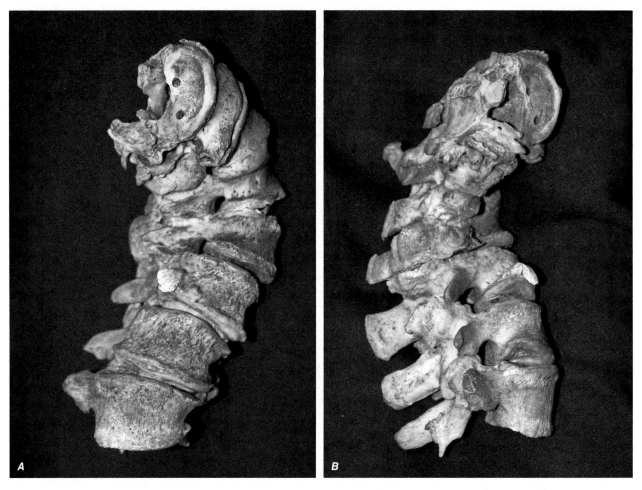

F I G U R E B-8.2.2. Contralateral hypoplasia: (*A*) anterior view, (*B*) right lateral view—T11 to L5 with right postlateral bar uniting T12 and L1 that are united with L2 and L3—all twisted into severe scoliosis, adult, NMNH anatomical collection.

adjacent vertebral bodies compensating with bony buttressing (Fig. B-9.4). The remaining dorsal portion can take on the appearance of an anterior–posterior hemivertebra (Fig. B-9.1D). Ventral hypoplasia can be mild or severe, often confused with compression fracture, particularly in the older individual as degenerative joint disease lesions develop. Ventral hypoplasia usually affects only one or two vertebral bodies, and most often appears in the lower thoracics, especially T12. The resulting spinal kyphosis can vary from mild to severe, depending on the degree of the defect (Epstein 1976; Schmorl and Junghanns 1971).

B-10. Dorsal Hypoplasia/Aplasia

This occurs with underdevelopment or absence of the dorsal cartilaginous center of the centrum with the ventral cartilaginous center unaffected as the embryonic vertebral body takes form. As the two centers quickly coalesce soon after they appear to form the future ossification center for the centrum, the dorsal portion lags behind the ventral portion, forming a dorsal wedge-shaped vertebral body (Fig. B-10). Dorsal hypoplasia generally appears in the lower lumbars, especially L5, and can vary from mild, to severe with spinal lordosis. Dorsal aplasia is extremely rare, forcing unopposed growth of the pedicles that fuse to the ventral portion of the vertebral body. Usually, cartilaginous tissue fills the void, preventing disturbance in spinal posture.

Extremely rare is the total agenesis of both dorsal and ventral embryonic cartilaginous centers, leading to the development of the neural arches without a centrum. The developing, unopposed pedicles grow around the spinal cord and fuse together to form a bony ring.

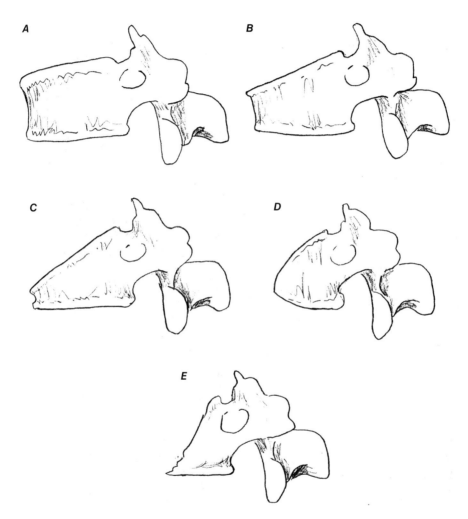

FIGURE B-9.1. Ventral hypoplasia/aplasia: (*A*) normal lower thoracic; (*B*) mild ventral hypoplasia; (*C*) severe ventral hypoplasia; (*D*) ventral aplasia with dorsal appearing hemivertebra; (*E*) ventral aplasia.

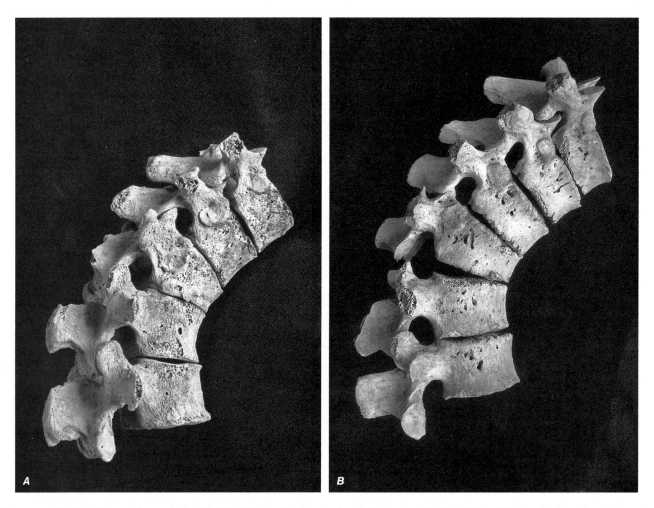

FIGURE B-9.2. Ventral hypoplasia: (*A*) T12 (T10 to L2 showing), adult female; (*B*) T12 (T9 to L2 showing), male adolescent, Bronze Age Da Shan Qian, Inner Mongolia, PRC.

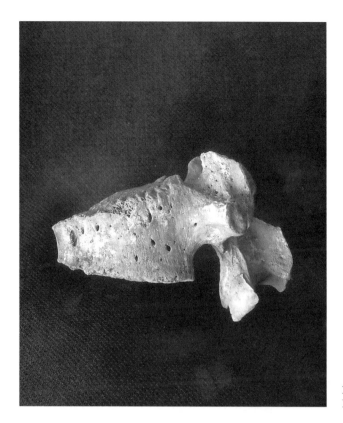

FIGURE B-9.3. Ventral hypoplasia: L1, male adolescent, Bronze Age Da Shan Qian, Inner Mongolia, PRC.

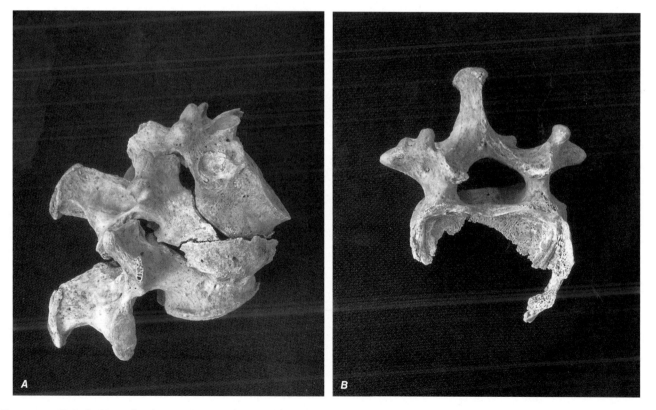

FIGURE B-9.4. Ventral aplasia: (*A*) L1 and ventral hypoplasia L2 with upward bony extension compensating for the absent anterior aspect of L1; (*B*) superior view of L1 ventral aplasia, adult female, Bronze Age Da Shan Qian, Inner Mongolia, PRC.

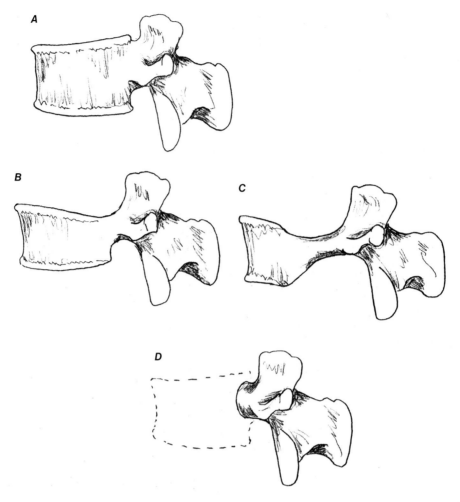

FIGURE B-10. Dorsal hypoplasia/aplasia: (*A*) normal lumbar vertebra; (*B*) mild dorsal hypoplasia; (*C*) severe dorsal hypoplasia; (*D*) complete aplasia of the centrum with pedicles joining to form a bony bar around the spinal canal.

B-11.1. Single Block Vertebra

Adjacent vertebral segments fail to separate when the fissure between precursor developing units of resegmented sclerotome does not appear (Fig. B-11.1.1). Complete unification occurs with total lack of the fissure, or partial unification when only part of the fissure is absent, with just the vertebral bodies united or just the neural arches united, or portions of both united. The isolated single conjoined block vertebra is usually not pathological as the joined vertebral segments maintain integrity with the same dimensional separation expected for the disk space between separated vertebral segments. This type of isolated single block vertebra is not uncommon, appearing as a familial trait, mostly within the cervical spine, especially with C2–C3 or C3–C4. Occasionally, it appears in the thoracic spine (Fig. B-11.1.2), especially near the border with the lumbars, but very rarely in the lumbar spine. Sometimes more than one single block vertebrae is present (Fig. B-11.1.3). Rarely, atypical malformed single block vertebra can appear that can be pathological (Fig. B-11.1.4).

B-11.2. Multiple Block Vertebra

Three or more developing vertebral segments fail to separate. Multiple blocks are often accompanied by other vertebral developmental disturbances, especially hemivertebrae, ventral and/or lateral hypoplasia, forming an atypical block vertebra. There is less of a chance for pathology if no other vertebral defect is present and the integrity of the individual vertebral segments within the block is maintained (Fig. B-11.2).

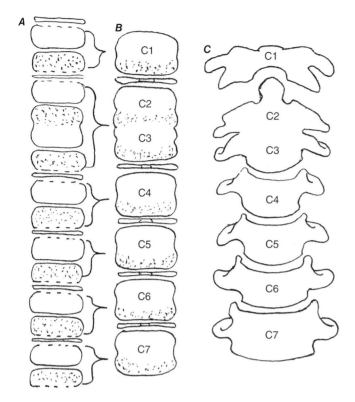

FIGURE B-11.1.1. Single block vertebra development: (*A*) failure of fissure formation between resegmenting sclerotomes for C2 and C3; (*B*) formation of C2–C3 block vertebral segments; (*C*) C2–C3 block vertebra.

FIGURE B-11.1.2. Single T3–T4 block vertebra: adult female, Byzantine Aphrodesias, Turkey.

B-11.3. Klippel–Feil Multiple Block Vertebra

Congenital brevicollis (shortened neck) applies to massive coalition of cervical vertebral segments into one multiple block vertebra (Fig. B-11.3.1) that may involve upper thoracic vertebral segments. Usually, the atlas is not included. The united vertebral segments do not maintain the normal integrity of unaffected vertebral segments, as ventral hypoplasia usually accompanies the coalescence. Other forms of vertebral disturbances may also be associated with this mass coalition. The defective cervical vertebral block produces an abnormal curvature of the cervical spine, shortening the neck and limiting its movement (Urunuela and Alvarez 1994a). This is classic K-L or more aptly called congenital brevicollis (Fig. B-11.3.2).

Klippel–Feil, first coined in the early part of the twentieth century as part of a syndrome and classified to include all types of block vertebrae (including isolated single blocks) into three types (Barnes 1994a:67), has always suffered from confusion as to how to place the many variable expressions of block vertebrae into this tight classification system, especially since single block vertebra and simple multiple block types do not lead to frank pathology. Classic K-L with the mass fusion of cervical vertebrae affected by ventral hypoplasia, with or without upper thoracic involvement, should rule as the definitive type in paleopathology. It may or may not be accompanied by other disturbances within the vertebral column, other parts of the skeleton, and/ or other organs.

B-12. Neural Arch Complex Disorders

These include hypoplasia/aplasia of the transverse processes, pedicles, laminae, spinous process, or apophyseal facets (Fig. B-12.1.1), excluding neural arch clefting (see Section B-3). Disruption of the development of one or both halves of any of the elements can lead to unilateral or bilateral insufficient ossification. Asymmetrical development based on unilateral hypoplasia without clefting often appears (Figs. B-12.1.1B and B-12.1.2),

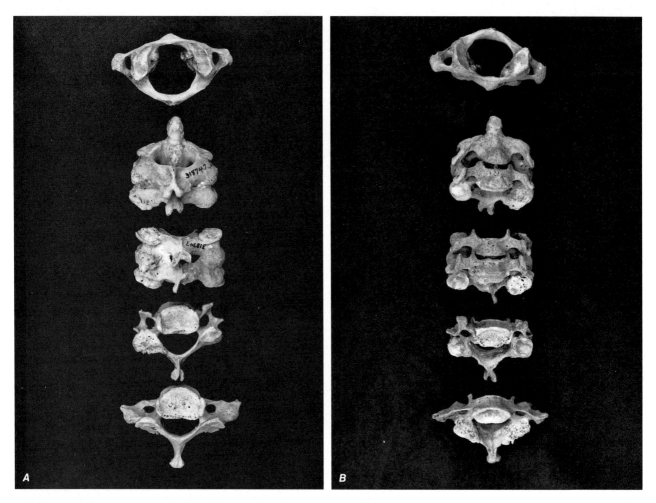

FIGURE B-11.1.3. Single C2–C3 and C4–C5 double block vertebrae: (C1–C7 showing) (*A*) dorsal view; (*B*) anterior view, adult male (NMNH 318747), anatomical collection.

with one or more posterior elements involved, and often leading to clinical symptoms. Asymmetrically formed neural arches within the lower lumbar spine are not uncommon, creating uneven load bearing that with overloading can lead to spondylolysis. Bilateral hypoplasia without clefting, especially in the atlas, results in a smaller than usual size posterior arch (Fig. B-12.1.1A), that is often overlooked and causes no clinical symptoms (Epstein 1976). Very rarely, part of the neural arch can be divided transversely (Fig. B-12.1.3).

Pedicle agenesis is rare (Fig. B-12.1.1C) and usually occurs unilaterally, most often between the fourth and seventh cervicals. Usher and Christensen (2000) identified a rare occurrence of a missing pedicle in the twelfth thoracic vertebra within multiple vertebral anomalies. Unless traumatized, the defect remains asymptomatic as the nonossified element remains as cartilaginous tissue. The associated apophyseal facet is usually missing, and the transverse process may also be absent or elongated and out of alignment (Wilson and Norrell 1966).

Apophyseal facets can be affected by hypoplasia, aplasia, or hyperplasia, either bilaterally or unilaterally, altering the development of counterpart facets. They can also be oriented in a transverse direction from the usual vertical position. They are usually affected when the laminae are underdeveloped, but apophyseal disturbance can appear separately (Fig. B-12.2.1A–C). Unilateral disturbances create uneven load bearing, leading to clinical symptoms. Asymmetrical apophyseal facets are common at the L-S border.

Transverse processes may not develop (aplasia), be rudimentary (hypoplasia), or enlarged (hyperplasia), either bilaterally or unilaterally (Figs. B-12.2.1D–F, B-12.2.2, and B-12.2.3). The costal portion of the cervical vertebral transverse elements may not ossify, leaving the transverse foramina to appear cleft or absent.

Spinous processes are often affected by hypoplasia, resulting in a rudimentary or shortened form (Figs. B-12.2.1G–I and B-12.2.4). The spinous process may not develop at all (aplasia). Enlarged spinous process

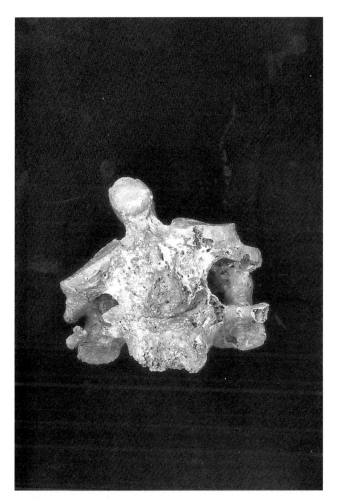

FIGURE B-11.1.4. Single C2–C3 block vertebra: asymmetrical from C3 left lateral hypoplasia, adult female, Byzantine Petras, Crete.

(hyperplasia) can also occur. The sacral crest, formed from rudimentary spinous processes, can also be affected by agenesis, and occasionally, hyperplasia creates an enlarged sacral crest. The last segment of the coccyx may be absent (aplasia) and often appears smaller than usual (hypoplasia).

B-13. Atlas Posterior/ Lateral Bridging

This may appear as a complete bony bridge or bony spicule (incomplete bony bridge) arching over the final pathway of the vertebral arteries bundled with veins and cervical nerves, passing upward through the cervical transverse foramina (Fig. B-13.1A). The final pass

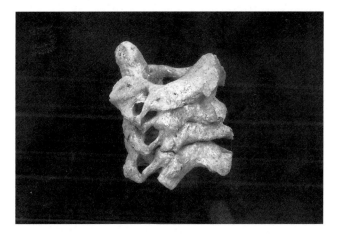

FIGURE B-11.2. Multiple block vertebrae: C2–C5, adolescent male, Frankish Corinth, Greece.

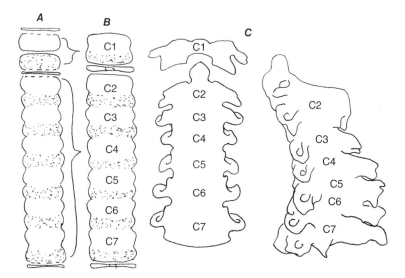

FIGURE B-11.3.1. Klippel–Feil (congenital brevicollis) development: (*A,B*) multiple failure of fissure formation between resegmenting cervical sclerotomes; (*C*) C2–C7 united with ventral hypoplasia.

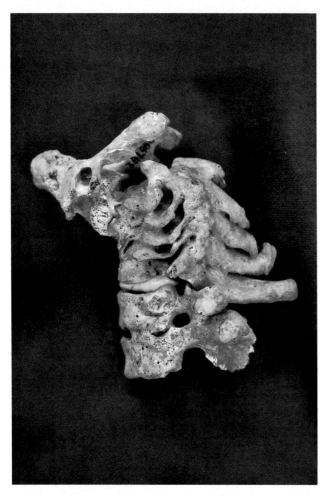

FIGURE B-11.3.2. Klippel-Feil (congenital brevicollis): C3–C7 multiple block with ventral hypoplasia and aplasia, C2 separate (unknown for C1) and T1–T2 single block, adult female (NMNH 382980), Plains Indian, USA.

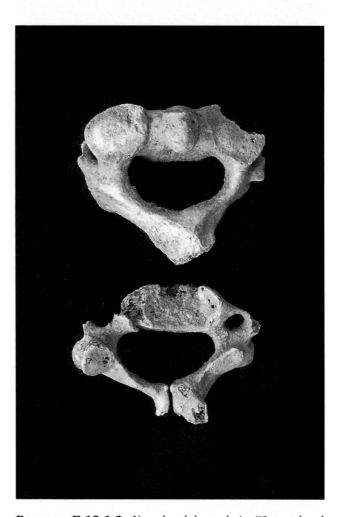

FIGURE B-12.1.2. Neural arch hypoplasia: C2 neural arch unilateral hypoplasia, C3 neural arch opposite unilateral hypoplasia with cleft, adult female, Classical Soukoulis cemetery, Corinth, Greece.

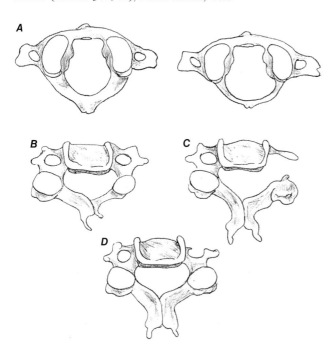

FIGURE B-12.1.1. Neural arch complex disorders: (*A*) normal atlas and bilateral hypoplasia of the posterior arch of the atlas; (*B*) asymmetrical neural arch (unilateral hypoplasia); (*C*) pedicle agenesis; (*D*) unilateral right agenesis cervical costal process.

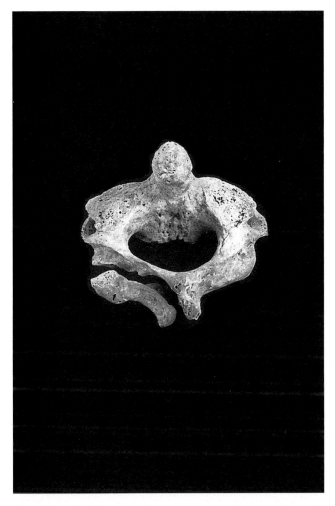

FIGURE B-12.1.3. Neural arch divided transversely: C2 unusual unilateral transverse separation of neural arch, child 8–9 years, Frankish Corinth, Greece.

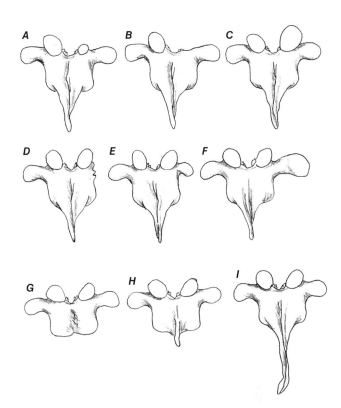

FIGURE B-12.2.1. Neural arch-associated disorders: (*A*) unilateral hypoplasia apophyseal facet; (*B*) unilateral apophyseal facet agenesis; (*C*) unilateral hyperplasia apophyseal facet; (*D*) unilateral agenesis transverse process; (*E*) unilateral rudimentary (hypoplasia) transverse process; (*F*) unilateral hyperplasia transverse process; (*G*) agenesis spinous process; (*H*) rudimentary (hypoplasia) spinous process; (*I*) hyperplasia spinous process.

is up through the transverse foramina of the atlas, coursing across the grooves in the superior lateral aspects of the first cervical before ascending into the foramen magnum. Posterior bony bridging can extend across the groove (sulcus) from front to back (Figs. B-13.1B,C,E and B-13.2), while the lateral bony bridging can cross where the artery leaves the transverse foramen to enter the groove (Fig B-13.1D,F). Posterior bridging is much more common than lateral bridging, and they usually appear unilaterally and are familial (Selby et al. 1955). Both bridging expressions very rarely develop together. This phenomenon appears to be a remnant of an evolutionary trend away from atlas bony bridging for the vertebral arteries (Le Minor and Trost 2004).

B-14. Multiple Vertebral Anomalies

These involve various types of developmental disturbances resulting from an array of disrupted genetic signaling within the developing vertebral column (Figs. B-14.1 and B-14.2). Klippel–Feil multiple vertebrae are often part of this phenomenon. Deciphering the various defects that appear can be challenging, as variable expressions of mixed vertebral segment disorders are rarely alike.

B-15. Sacral Agenesis versus Hemisacrum

Sacral agenesis is part of the caudal regression syndrome. During gastrulation as the primitive streak regresses caudally, the caudal eminence develops from a mass of mesodermal cells to form the tail end of the embryo. Mesodermal cells within the eminence form a solid neural cord that extends to meet the caudal end of the developing neural tube that usually ends at the level of the developing second sacral segment. By a process of secondary neurulation, the neural cord

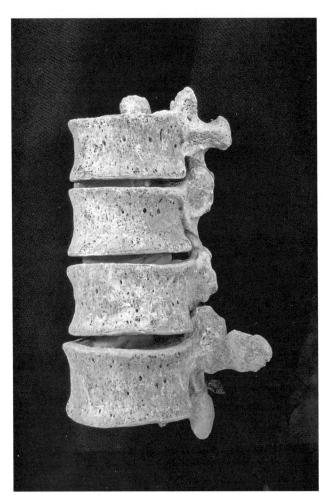

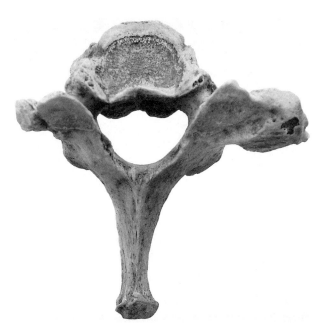

FIGURE B-12.2.3. Transverse process hypoplasia: T1, adult, unilateral left hypoplasia transverse process (contributed by C.F.Merbs).

FIGURE B-12.2.2. Transverse process aplasia: T12 and L1 bilateral aplasia transverse processes (shown with T11 and L2), adult female, Frankish Corinth, Greece.

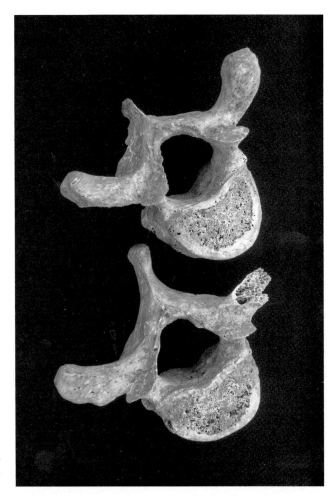

FIGURE B-12.2.4. Spinous process hypoplasia: T2 hypoplasia —remnant spinous process with normal T3 below, adult female, Frankish Corinth, Greece.

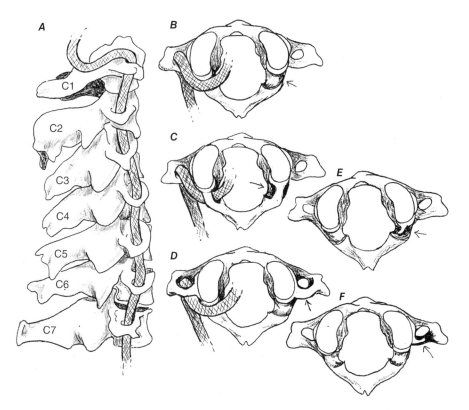

FIGURE B-13.1. Atlas posterior/lateral bridging: (*A*) right vertebral artery ascending through the cervical transverse foramina, crossing over atlas groove; (*B*) arrow pointing to the groove without artery present on the left side; (*C*) arrow pointing to the posterior bridge over the groove on the left side (right side shows artery passing under the bridge); (*D*) arrow pointing to the lateral bridge on the left side (right side shows artery passing under the bridge before entering the groove); (*E*) arrow points to the incomplete posterior bridge; (*F*) arrow points to the incomplete lateral bridge.

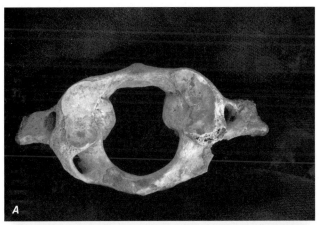

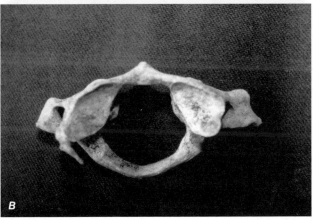

FIGURE B-13.2. Atlas posterior bridging: (*A*) unilateral complete, adult male, La Playa, NW Mexico; (*B*) unilateral incomplete, adult male (NMNH 308613), Hawikuh, NM.

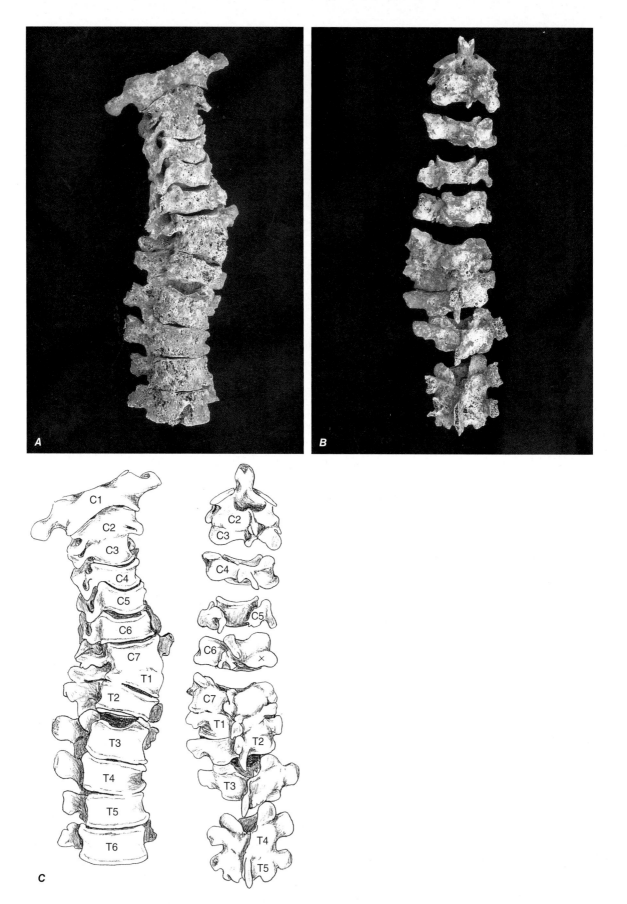

FIGURE B-14.1. Multiple vertebral anomalies: (*A*) anterior C1 to T6—normal atlas, single blocks C2–C3 and T4–T5, wide cleft neural arch C5, extra neural arch segment joined to C6, multiple block C7-T1-T2 with severe right lateral hypoplasia T1, mild lateral hypoplasia C4, C7, T2, T3, and T4, normal T6; (*B*) dorsal C2 to T5; (*C*) drawings of *A* and *B*, adult male, Ottoman Corinth, Greece.

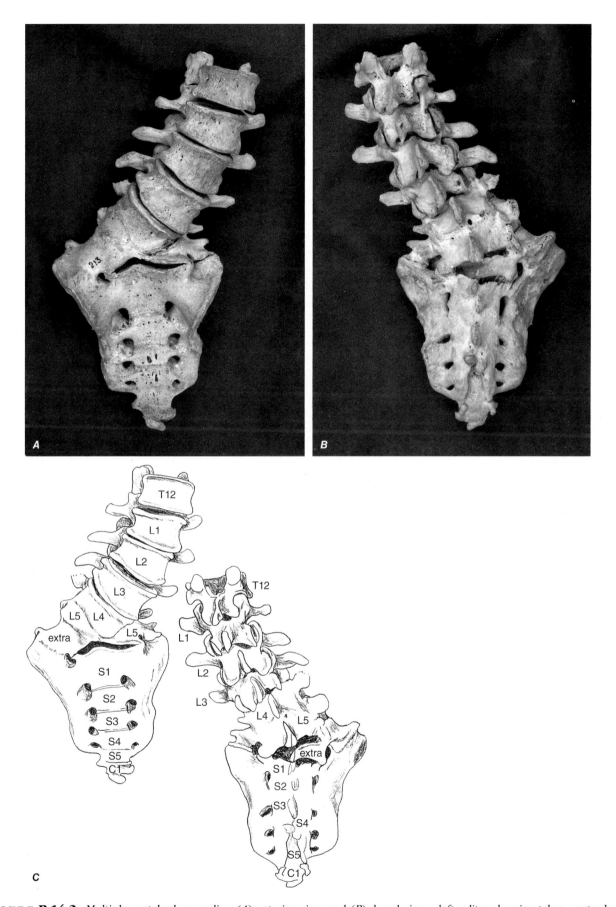

FIGURE B-14.2. Multiple vertebral anomalies: (*A*) anterior view and (*B*) dorsal view—left solitary hemivertebra—extra hemi-metamere segment united with the sacrum, sagittal cleft L5 united with L4 in the block vertebra incompletely united to the sacrum, lateral hypoplasia L3, L2, and L1, normal T12, and sacralized caudal segment; (*C*) drawings of *A* and *B*, adult male (213), NMNH Terry collection.

undergoes cavitation as it connects to the caudal end of the neural tube to house the filum terminale. This process also facilitates the development of the most caudal somites of the sacrum and coccyx (Larsen 2001). Caudal regression infers that the development of the caudal eminence is disturbed, leading to a wide spectrum of developmental disorders, from minor to major, that can affect not only the osseous structures that form the sacrum and coccyx, but also the developing bladder, urethra, lower bowel, and nerve innovation to these organs, and affect the development of the lower limbs. Severe cases affect the lower limbs with hypoplasia, especially the femora, with rigid outward flexion of the hips and knees, along with foot deformities. The most extreme disturbance leads to sirenomelia (fused limb buds). Severe forms of caudal regression can also interfere with the development of the lower lumbar spine, with very severe disturbances extending upward to involve the lower thoracic spine (Larsen 2001).

Sacral agenesis can be confused with varying expressions of sacral–coccygeal somite defects—unilateral hypoplasia/aplasia, or the formation of hemivertebrae, especially when several sacral segments are involved. I previously confused sacral agenesis with hemisacrum—unilateral aplasia/hypoplasia of all sacral segments (Fig. B-15.1). Somite defects often accompany expressions of sacral agenesis, adding to the confusion.

Complete sacral agenesis is virtually unknown in paleopathology. Partial sacral agenesis, the most likely form to be identified in human skeletal remains, leaves the uppermost sacral segments in place, either S1 alone or S1–S2, or missing only the lower one or two segments and the coccyx, or just the coccygeal segments (Fig. B-15.2). Neurological disturbance, especially motor nervation, can vary, depending on the level of disturbance in development. Generally, caudal neurulation outpaces bony structural development so that neurological impairment with sacral agenesis can vary from mild to severe. With S1 or S1–S2 segments in place, articulating with the ilia, the pelvis will not suffer distortion and neurological deficit will be mild, allowing the affected individual to survive into adulthood in the past. If all sacral segments are absent, the last lumbar may articulate with the ilia, or rest above the ilia as they meet midline to articulate or fuse together, with greater

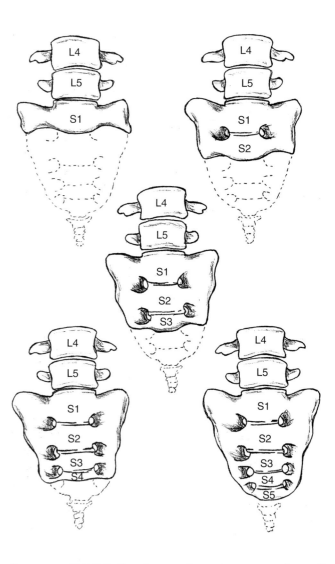

FIGURE B-15.2. Sacral agenesis: various expressions of partial agenesis from the absence of S1 to missing only the coccyx.

FIGURE B-15.1. Hemisacrum: unilateral aplasia/hypoplasia of all sacral segments.

neurological disturbance, leaving the affected individual little chance of surviving early infancy in the past. Most likely, milder forms missing only one or two lower sacral segments along with the coccyx, or just missing the coccyx, having little impact on the affected individual, will be found in past human skeletal remains.

B-16. Enlarged Anterior Basivertebral Foramina

This appears as enlarged anterior grooves on the thoracic and lumbar vertebral bodies of some children. The grooves are formed by the convergence of tortuous blood vessel channels forming large vascular pools that intrude onto the anterior surface of the developing vertebral bodies. The enlarged foramina are lined with smooth cortical-like bone within the trabecular bone tissue of the vertebral bodies (Figs. B-16.1 and B-16.2). As the child grows, these enlarged vascular channels usually decrease in size with much smaller appearing anterior foramina but they occasionally persist into adolescence. They are rarely found in adults and appear not to be pathological (Ferguson 1968). Some medical researchers believe that persistent enlarged anterior basivertebral foramina of midthoracic vertebrae leads to adolescent kyphosis, but most likely, the kyphosis results from ventral hypoplasia of the vertebral bodies.

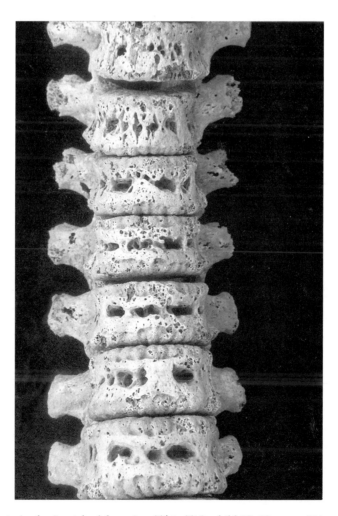

FIGURE B-16.1. Enlarged anterior basivertebral foramina: T4 to T10, child 12–13 years, Ottoman Corinth, Greece.

FIGURE B-16.2. Enlarged anterior basivertebral foramina: (*A*) young child; (*B*) early adolescence; (*C*) young adult.

RIBS

RIB DEVELOPMENT

The embryonic sclerotome tissue forming the dorsal–lateral aspects of the vertebral column contributes to the formation of the costal portions of the transverse processes of all the vertebral segments (Fig. C-1.0). The costal portions of the 12 thoracic vertebrae receive genetic signals to separate and extend forward and downward into the formation of bilateral ribs to house the thorax region. Cartilage extensions from the bony rib ends reach forward from the first seven pairs of developing ribs to attach to the developing sternum. Cartilage extensions for the eighth and ninth ribs do not join the sternum but join together and attach to the cartilage extension of the seventh rib. The last two pairs remain free of costal cartilage ends. The articular head ends interface with the thoracic bodies, while facets interface with the transverse processes for the first 10 ribs; the last two ribs articulate only with the vertebral bodies. Secondary ossification of cartilaginous epiphyses for the head, articular and nonarticular parts of the tubercles, begins in late puberty, uniting with the rib body by early adulthood. Rib length increases from the first rib to the seventh or eighth, then gradually decreases until the last rib that is usually very short but variable in length. The costal portions of other vertebrae are usually programmed to loose their potential to form ribs.

Developing ribs can be affected by irregular segmentation of the costal precursors. Rib development and direction can also be disrupted by abnormal development of the thoracic vertebrae, forming a distorted thorax.

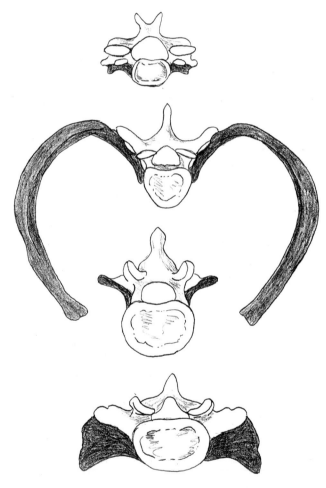

FIGURE C-1.0. Rib development: from shaded costal portions of the thoracic vertebra compared with the shaded costal portions of the cervical vertebra (top), lumbar vertebra (below the thoracic vertebra with ribs), and sacral vertebral segment (bottom).

Atlas of Developmental Field Anomalies of the Human Skeleton: A Paleopathology Perspective, First Edition. Ethne Barnes.
© 2012 Wiley-Blackwell. Published 2012 by John Wiley & Sons, Inc.

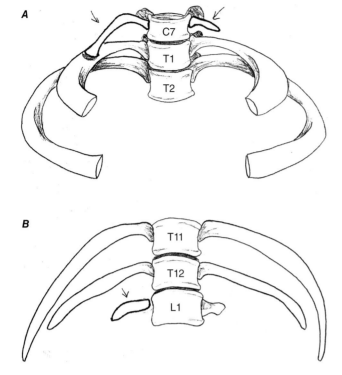

FIGURE C-1.1. Supernumerary ribs: (*A*) cervical ribs from cranial shifting of the cervical-thoracic vertebral border; (*B*) lumbar rib from caudal shifting of the thoracic–lumbar vertebral border.

FIGURE C-1.2. Supernumerary intrathoracic rib: (*A*) parallel to the parent rib; (*B*) oblique to the parent rib.

RIB ANOMALIES

C-1. Supernumerary Ribs

These appear with vertebral border shifting. Shifting cervical–thoracic or thoracic–lumbar vertebral borders can trigger affected vertebral costal portions to develop into ribs or rib-like extensions (Fig. C-1.1). Cranial shifting of the cervical–thoracic border can force the seventh cervical vertebra to create a cervical rib (Section B-1.1), while caudal border shifting of the thoracic–lumbar border may force the first lumbar vertebra to create a lumbar rib (Section B-1.4). The appearance of rib or rib-like formations developing from costal portions of all other vertebral segments is extremely rare.

Transitional Vertebra Extra Rib
Rudimentary ribs often accompany transitional vertebral segment taking on the characteristics of a thoracic vertebra.

Intrathoracic Rib
This appears on rare occasions without causing pathology, usually forming unilaterally on the right side of the

middle thorax (Guttenag and Salwen 1999). The embryonic vertebral costal portion separates, creating two distinct developing rib portions from a single parent entity. The smaller, additional rib segment developing from the inferior–posterior border of the parent rib either grows parallel to it, or takes an oblique descending position (Fig. C-1.2) that may have a fibrous attachment to the diaphragm with some interference with full ventilation.

C-2. Rib Hypoplasia/Aplasia

Primarily affects the first and last ribs with vertebral border shifting (Fig. C-2). Caudal border shifting of the cervical–thoracic border can reduce the size of the first rib to rudimentary form (Section B-1.2), while cranial

A

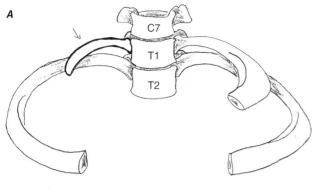

B

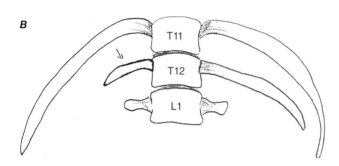

FIGURE C-2. Rib hypoplasia: (*A*) rudimentary first rib from caudal shifting of the cervical–thoracic vertebral border; (*B*) rudimentary twelfth rib from cranial shifting of the thoracic–lumbar vertebral border.

A

B

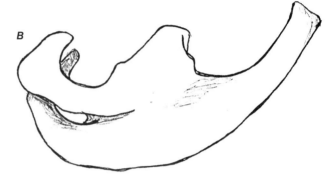

FIGURE C-3.1. Merged ribs: (*A*) cervical rib merged with the first rib; (*B*) first rib merged with the second rib.

border shifting at the thoracic–lumbar border can greatly reduce the size of the twelfth rib, or eliminate it all together (Section B1.3). Reduction in size or absence of other ribs is extremely rare.

C-3. Merged Ribs

This anomaly usually affects the first two ribs, or if a cervical rib is present, it can merge with the first rib (Fig. C-3.1). The merging of the two ribs in development leaves them united as if fused together. Other ribs may merge or fuse together (Fig. C-3.2), especially at the vertebral ends (Fig. C-4), especially when counterpart vertebral segments are disturbed with irregular segmentation.

C-4. Bifurcated Ribs

This anomaly usually occurs in the right third, fourth, or fifth rib, appearing forked as the sternal ends separate during development (Fig. C-4).

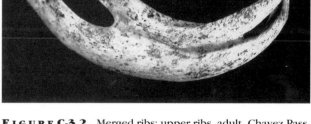

FIGURE C-3.2. Merged ribs: upper ribs, adult, Chavez Pass, AZ (contributed by C.F. Merbs).

C-5. Other Rib Disorders

Bridged Ribs

This disorder is closely linked to merged ribs as a bony bridge forms between adjacent ribs (Fig. C-5.1A). Sometimes an incomplete articulating bridge forms (Figs. C-5.1B and C-5.2).

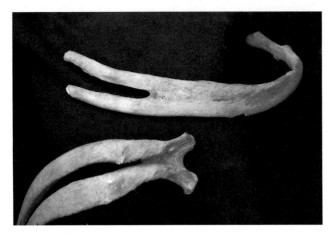

FIGURE C-4. Bifurcated and merged ribs: bifurcated rib and merged vertebral end of the rib, adults, NMNH anatomical collection.

FIGURE C-5.2. Incomplete bridged ribs: incomplete articulating bony bridge, adult male, Ottoman Corinth, Greece.

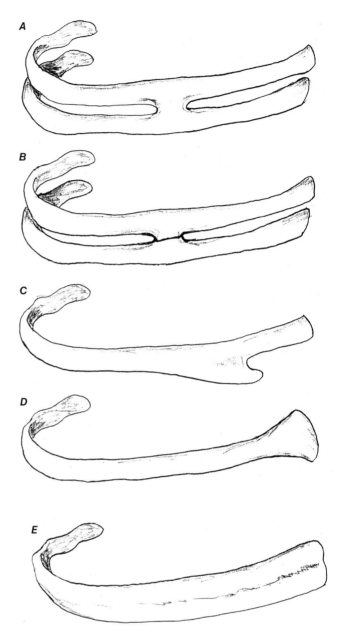

A

B

C

D

E

FIGURE C-5.1. Other rib disorders: (*A*) complete bony bridge; (*B*) incomplete articulating bony bridge; (*C*) rib spur; (*D*) flared sternal rib end; (*E*) hyperplasia—extra wide rib.

Rib Spur
Sometimes a large bony spur develops near the sternal end, appearing much like an incomplete bifurcated end (Fig. C-5.1C).

Flared Rib
The sternal end may appear much wider than the rib body (Fig. C-5.1D).

Rib Hyperplasia
This appears as if two ribs were conjoined, with its width equal to or exceeding that of two normal ribs, resulting in an extra wide rib (Fig. C-5.1E).

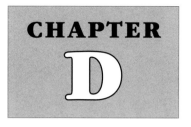

CHAPTER D

STERNUM

STERNUM DEVELOPMENT

Embryonic development of the sternum is complex, derived from a pair of mesenchymal tissue condensation bands or bars forming ventrolaterally ahead of developing membranous rib segments (Fig. D-1.0). Moving toward the midline ahead of the upper primordial rib segments, the sternal bands first meet at their cranial ends, where they are joined by another mesenchymal condensation, the precostal process, to form the precursor manubrium. Each side of the superior border of the developing primordial manubrium is then joined by one of a pair of small mesenchymal condensations known as the suprasternal structures that will form the bony interface between the manubrium and clavicles. The remaining portions of the sternal bands follow a cranial–caudal gradient as they reach midline and join together ahead of the developing rib segments. Once the sternal bands have joined and the cartilage model for the sternum begins to take shape, the fused sternal bands below the manubrium separate into four segments known as sternebrae, with residual caudal tissue becoming the precursor xiphoid process. The manubrium and segmented sternebrae usually present at birth as separate bones, while the xiphoid process remains cartilaginous until a few years later. The cartilage separating the sternebrae ossify and unite them as one in a caudal–cranial direction beginning in puberty with completed union by the mid-20s, while the manubrium generally remains separated from the mesosternum by a fibrous lamina. The xiphoid process similarly may also remain separated from the caudal end of the mesosternum. Adult male sterna are generally larger than female sterna. Variations do occur in the timing and sequence of the sternebrae fusion, the number of sternebrae, size of the xiphoid process, and shape of the mesosternum. Early fusion of two or more sternebrae and early ossification of the xiphoid can be seen as early as infancy.

STERNUM ANOMALIES AND VARIATIONS

D-1. Suprasternal Ossicles

Sometimes one or both embryonic suprasternal structures fail to meet their designated connection between the developing sternal ends of the clavicles and the evolving membranous manubrium (Fig. D-1.1). This displacement of the suprasternal structures, usually above or near the suprasternal (jugular) notch, does not appear to prevent the development of functional sternoclavicular joints. Left unattached, they develop into separate cartilaginous structures that ossify into bony nodules during adolescence, varying in size from 2 to 15 mm. The ossicles may fuse together or remain separate. They may fuse to the manubrium in or near the suprasternal (jugular) notch (Fig. D-1.2), or remain as separate ossicles within the interclavicular ligament that

Atlas of Developmental Field Anomalies of the Human Skeleton: A Paleopathology Perspective, First Edition. Ethne Barnes.
© 2012 Wiley-Blackwell. Published 2012 by John Wiley & Sons, Inc.

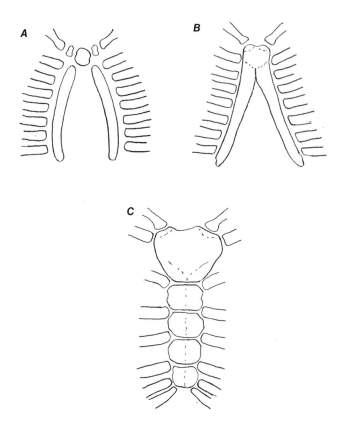

FIGURE D-1.0. Sternum development: (*A*) sternal bands moving to the midline ahead of the upper seven rib segments, with cranial portions meeting the precostal process and paired suprasternal structures moving ahead of the developing clavicles; (*B*) fusion of the upper cranial portions of the sternal bands with the precostal process and suprasternal structures to form the manubrium; (*C*) final development of the immature sternum—manubrium and four sternebrae segments with attached seven paired costal cartilages and articulating clavicles.

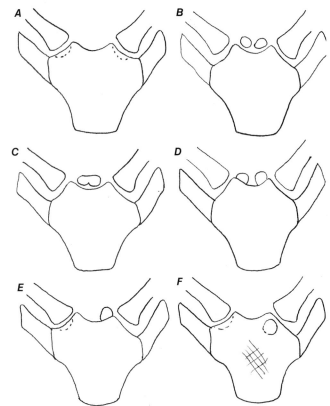

FIGURE D-1.1. Suprasternal ossicles: (*A*) manubrium with sternal ends of clavicles and first costal cartridges (dotted lines represent suprasternal structures); (*B*) bilateral suprasternal ossicles; (*C*) conjoined suprasternal ossicles; (*D*) bilateral suprasternal ossicles fused to the manubrium suprasternal notch; (*E*) unilateral suprasternal ossicle fused with the manubrium suprasternal notch; (*F*) unilateral suprasternal ossicle fused to the dorsal aspect of the manubrium.

crosses over the suprasternal notch of the manubrium, attaching to the superior sternal ends of the clavicles. They can be unilateral or bilateral, with the majority bilateral separate ossicles (Stark et al. 1987). Separate ossicles are usually lost in archaeological recovery, leaving only those fused to the manubrium for identification.

D-2. Mesosternum Shape Variations

These depend on the timing of the sternal bands joining together in ascending order that in turn determines the number of ossification centers to appear later in each sternebra. Usually, the first portion below the manubrium forming the first and second sternebrae produces a single ossification center each, while the last two segments usually produce two ossification

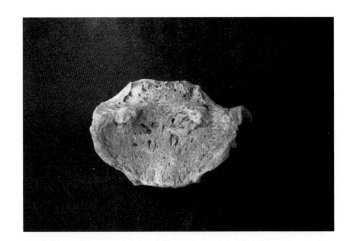

FIGURE D-1.2. Suprasternal ossicles: bilateral, fused to the dorsal aspect of the manubrium, adult male, Mishongovi, AZ (Field Museum).

centers each. Delay in the joining of the sternal bands or portions of it can lead to divided developing sternebrae with separated ossification centers. Divided sternebrae can be seen at birth, but they usually coalesce by early childhood. Thus, variation occurs in the shape of the mesosternum, particularly in the lower two sternebrae segments and also in the xiphoid process. Variations in shape basically range from a narrow mesosternum (type I), narrow cranial portion with wide caudal portion (type II), wide upper and lower portions (type III), to a wide upper portion with narrower lower portion (type IV) (Ashley 1956) (Fig. D-2.1). The last sternal segment is most often smaller than the others, and types I and II are the most common shapes to appear (Figs. D-2.2–D-2.5), while type IV is rare in humans.

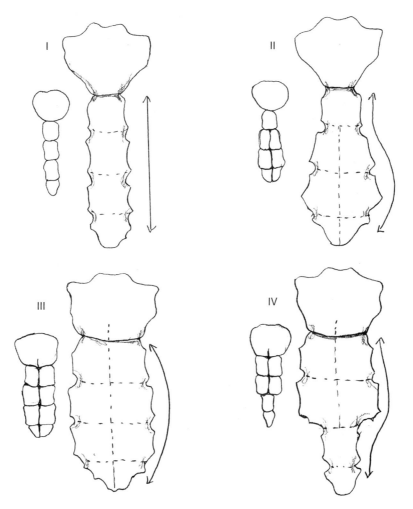

FIGURE D-2.1. Mesosternum basic variations: (infant and adult) type I narrow upper and lower portions; type II narrow upper portion, wider lower portion; type III wide upper and lower portions; type IV (very rare) wide upper portion, narrow lower portion.

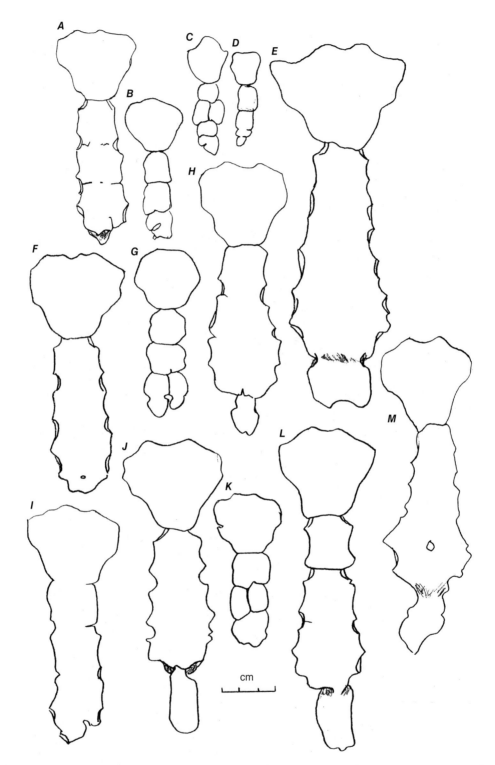

FIGURE D-2.2. Sternum variations—Greek: (*A*) adolescent 15-16 years; (*B*) child 6-7 years; (*C*) child 5.5-6.5 years; (*D*) infant 8-9 months; (*E*) adult male; (*F*) adult female; (*G*) child 7.5-9 years; (*H*) adult female; (*I*) adolescent female 17-18 years; (*J*) adult male; (*K*) child 7-8 years; (*L*) adult female; (*M*) adult female, Medieval Corinth, Greece (all drawn to same scale).

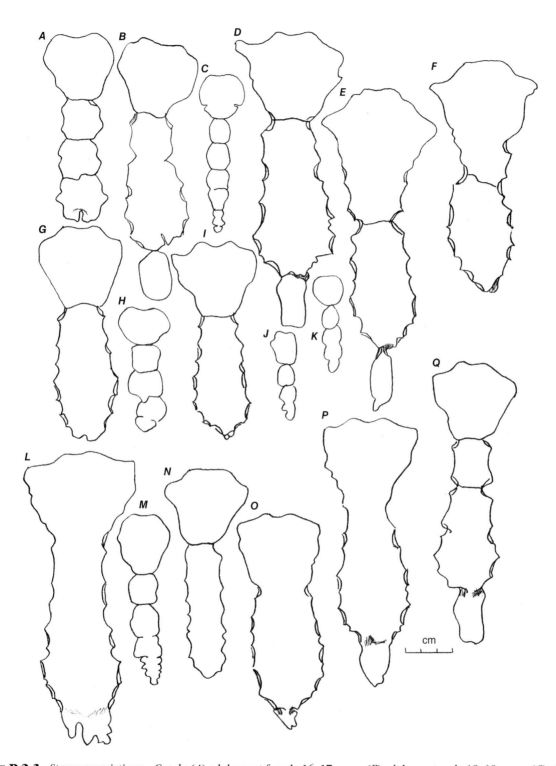

FIGURE D-2.3. Sternum variations—Greek: (*A*) adolescent female 16-17 years; (*B*) adolescent male 18-19 years; (*C*) child 3.5-4 years; (*D*) adult male; (*E*) adult male; (*F*) adult; (*G*) adult female; (*H*) child 5.5-6.5 years; (*I*) adult female; (*J*) infant 16 months; (*K*) infant 9-10 months; (*L*) adult male; (*M*) child 9-10 years; (*N*) adult female; (*O*) adult female; (*P*) adult female; (*Q*) adult female, Medieval Corinth, Greece (all drawn to same scale).

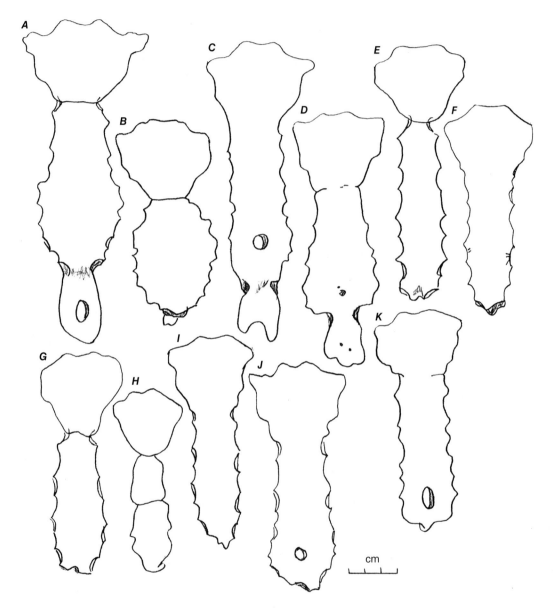

FIGURE D-2.4. Sternum variations—SW USA: (*A*) adult male; (*B*) adult female; (*C*) adult male; (*D*) adult female; (*E*) adult male; (*F*) adult female; (*G*) adult female; (*H*) child 11–12 years; (*I*) adult female; (*J*) adult female; (*K*) adult female, American Southwest (all drawn to same scale).

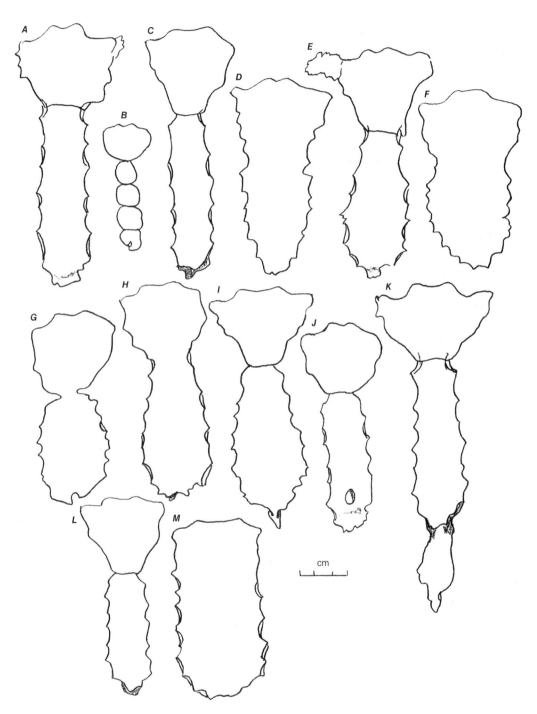

FIGURE D-2.5. Sternum variations—SW USA: (*A*) adult female; (*B*) child 7–8 years; (*C*) adult female; (*D*) adult female; (*E*) adult; (*F*) adult female; (*G*) adult female; (*H*) adult female; (*I*) adult female; (*J*) adult female; (*K*) adult male; (*L*) adult female; (*M*) adult male, American Southwest (all drawn to same scale).

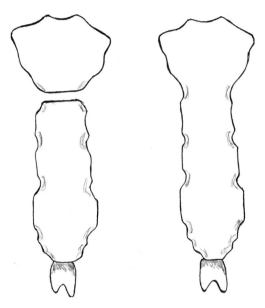

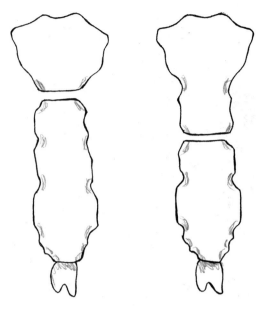

FIGURE D-3. Manubrium–mesosternal joint fusion: fibrous joint in place with separated manubrium compared with the absence of fibrous joint with manubrium fused to the mesosternum.

FIGURE D-4.1. Misplaced manubrium–mesosternal joint: fibrous joint between the manubrium and mesosternum compared with the fibrous joint misplaced between the first and second sternebrae.

D-3. Manubrium–Mesosternal Joint Fusion

This occurs when the fibrous lamina fails to develop between the developing manubrium and first sternebra of the mesosternum. Instead, the manubrium–mesosternal joint develops as a cartilaginous separation, similar to those formed between the sternebrae. As the sternebrae coalesce into the mature sternum, the manubrium also fuses with the mesosternum with failure of the fibrous lamina to develop (Fig. D-3). The xiphoid has a similar fibrous lamina separating it from the caudal end of the mesosternum, but often, it forms a cartilaginous separation that ossifies with maturity of the sternum. Usually, manubrium–mesosternal joint fusion does not appear until adolescence as the sternal segments unite, but occasionally, it can be seen as young as infancy (Ashley 1954).

D-4. Misplaced Manubrium–Mesosternal Joint

This develops when the fibrous lamina is misplaced between the first and second sternebrae, instead of between the evolving manubrium and first sternebra (Figs. D-4.1–D-4.3). The manubrium and first sternebra share a cartilaginous space leading to ossification that unites them as one with the intended manubrium–mesosternal joint below (Ashley 1954).

FIGURE D-4.2. Misplaced manubrium–mesosternal joint: adult male, Frankish Corinth, Greece.

FIGURE D-4.3. Misplaced manubrium–mesosternal joint with failed union of the mesosternum: adult male, Byzantine Petras, Crete.

D-5. Mesosternal Hypoplasia/Aplasia

This commonly affects the last sternebra segment and xiphoid process (Fig. D-5). Both bilateral and unilateral expressions can occur as one or both developing sternal bars are affected, resulting in a short or asymmetrical mesosternum. The xiphoid process can be absent or very small.

D-6. Sternal Hyperplasia

Extra long sternal bands produce an additional sternebra segment, forming an unusually long mesosternum. Sometimes, the extra sternal band tissue forms an extra long xiphoid process (Fig. D-6).

D-7. Sternal Aperture

This is often mislabeled as a sternal foramen. This aperture results from incomplete cohesion of a portion of

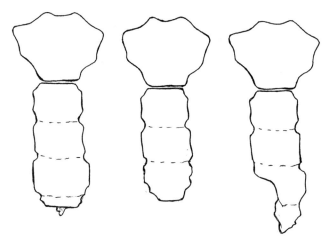

FIGURE D-5. Mesosternum hypoplasia/aplasia: left to right—hypoplasia last sternebra segment, aplasia last sternebra segment, unilateral aplasia last two sternebra segments.

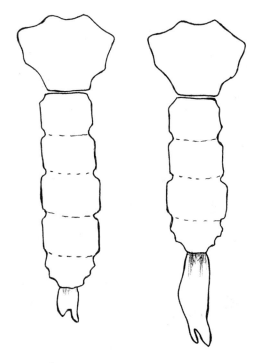

FIGURE D-6. Sternum hyperplasia: left to right—extra sternebra segment and extra long xiphoid.

the lower end of the embryonic sternal bands, generally between the third and fourth sternebra segments. The bony opening can be an oval or elongated aperture, varying in size according to the rift between the two sternal bars. A similar aperture can appear in the xiphoid process for the same reason (Fig. D-7). Fibro-cartilaginous tissue fills the void in place of the osseous

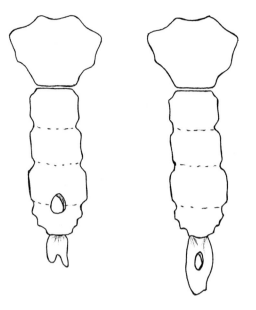

FIGURE D-7. Sternal aperture: mesostenal aperture and xiphoid process aperture.

tissue, eliminating the idea of a foramen (Cooper et al. 1988).

D-8. Sternal Caudal Clefting

This results from the caudal ends of the sternal bands too slow meeting at the midline. Usually, the delay is minor with only a cleft or fissure at the caudal border of the last sternebra segment. Sometimes, only the xiphoid process is cleft. Very rarely, a wide cleft forms when the caudal portion of the sternal bands fails to meet midline (Eijgelaar and Bijtel 1970) (Fig. D-8).

D-9. Bifurcated Sternum

A very rare cranial clefting as the sternal bands are unable to meet or are slow reaching the midline during morphogenesis (Fig. D-9). This may be influenced by a divided precostal mesenchymal condensation that fails to coalesce into one. Complete bifurcation divides the sternum from the top to bottom, but usually connecting to an intact xiphoid process. Fibrocartilaginous tissue fills the void and holds the divided segments together beneath the skin covering. The defect causes paradoxical motions that hamper cardiopulmonary functions with difficult breathing, attacks of cyanosis, and recurrent respiratory infections. The heart may be pushed outward from the chest cavity (ectopic cordis) with severe expressions of bifurcation, especially when

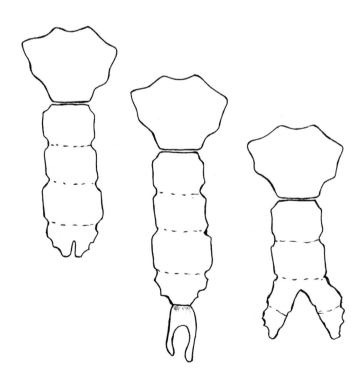

FIGURE D-8. Sternal caudal clefting: left to right—distal mesosternal cleft, xiphoid process cleft, and wide mesosternal cleft.

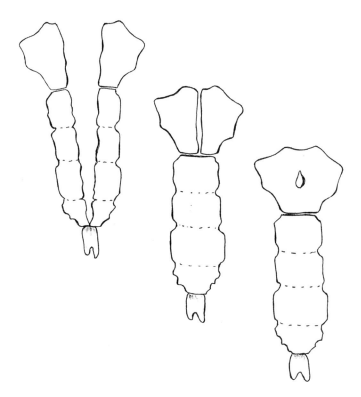

FIGURE D-9. Bifurcated sternum: left to right—complete cleft, bifurcated manubrium, and manubrium aperture.

the xiphoid process is absent, leaving the divided sternum without a bony anchor (often mislabeled as absent sternum). Partial bifurcation affects only the manubrium, either dividing into halves, or on a minor scale, forming an aperture in the middle of the manubrium (Eijgelaar and Bijtel 1970).

D-10. Pectus Excavatum (Funnel Chest)

This is present when growth of the mesosternum is misdirected inward generally by anomalous shortened growth and development of the central tendon of the diaphragm that usually attaches to the dorsal side of the xiphoid process. Shortened muscular tendon fibers attached to the lateral lower portions of the lower mesosternum may also pull the lower portion inward. Bony growth of the sternum itself is not affected, only the direction of growth. The depressed portion of the mesosternum can be mild to severe (Fig. D-10). Mild forms are asymptomatic, often affecting only the xiphoid and may go unnoticed in paleopathology. Sometimes, the xiphoid can be so depressed that it touches the vertebral column. Severe expressions are marked by deep depression of the lower one-third of the mesosternum and may affect respiration and displace the heart (Warkany 1971).

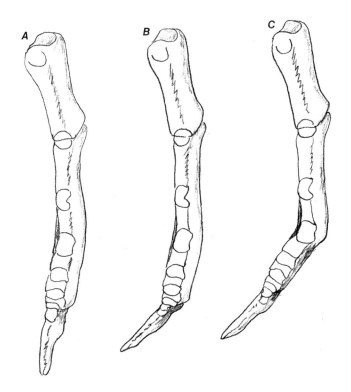

FIGURE D-10. Pectus excavatum: lateral contour—(A) normal sternum; (B) depressed xiphoid process; (C) depressed lower one-third.

D-11. Pectus Carinatum (Pigeon Breast)

This is present when the growth of the mesosternum is misdirected forward by anomalous growth and development of associated ribs and attached costocartilages. Bony growth of the sternum is not disturbed as the affected portion of the mesosternum is pushed forward. Bowing of the mesosternum, known as arcuate sternum, is a mild, asymptomatic form of this disorder that generally follows familial lines (Fig. D-11). The chest wall protrudes with enlarged sagittal diameter of the thorax. More drastic forms of this defect show a very pronounced forward arching of the upper or midportion of the mesosternum that can interfere with respiration (Warkany 1971).

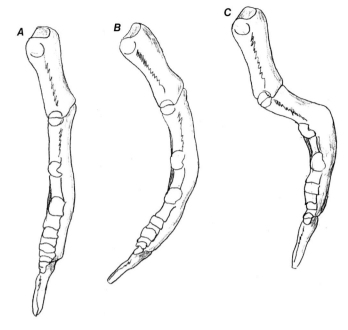

FIGURE D-11. Pectus carinatum: lateral contour—(*A*) normal sternum; (*B*) arcuate (bowed); (*C*) upper mesosternum pushed forward.

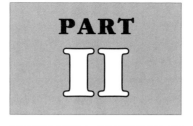

APPENDICULAR SKELETON

CHAPTER E

UPPER LIMBS

UPPER LIMB DEVELOPMENT

Limb morphogenesis occurs between the fourth and eighth weeks with the final pattern of development of the appendicular skeleton laid down by the end of the embryonic period (Fig. E-1.0). During the third week of development, bilateral swellings for the upper limb buds appear along the ventral aspect of the lateral mesoderm opposite the cervical–thoracic region of the 2.5-mm embryo. Budding is stimulated by a ring of overlying thickened ectoderm known as the crest of Wolff. The eruptions consist of undifferentiated mesenchymal cells from the lateral mesoderm, covered by a layer of ectodermal tissue. Ectoderm along the ventral–medial borders of the buds thickens to form apical ectodermal ridges (AERs), the primary movers for limb development. Underneath this oblong, thickened margin, the closest undifferentiated mesenchymal cells to the AER form the progressive zone (PZ) from which mesenchymal cells move in a proximal direction of the growing limb bud. Together they form a functional unit with a feedback system of genetic signals responsible for the outgrowth of the limb's mesenchymal core along its proximal–distal axis.

Cascading, overlapping genetic signals move along the proximal–distal gradient of the limb core as it lengthens distally. At designated positions along this pathway, beginning at the most proximal end of the shoulder girdle, differing combinations of overlapping genetic signals relay to the undifferentiated mesenchy-mal core in their pathway to activate specific formations of the different limb segments. The developing limb bud also responds to genetic signals from two other axes that contribute to positioning and patterning of limb formation through complex interactions with the proximal–distal axis. The major focus of the anterior–posterior axis is the development of the preaxial (radial—thumb side) and postaxial (ulnar—fifth finger side) aspects of the developing bud. Patterning along this axis is regulated by a cluster of cells at the posterior border near the flank of the limb bud adjacent to the AER, known as the zone of polarizing activity (ZPA). The anterior–posterior axis is responsible for the positioning of the digits in proper order as well as positioning of the forearm bones. Dorsoventral polarity is maintained in the developing limb bud by genetic signals restricted within the dorsal and ventral limb bud ectoderm. As the different components of the limb segments are defined, thickening of the mesenchymal cells between bone anlages occurs under genetic signaling. The mesenchymal cells within these areas, known as interzones, differentiate into fibrous connective tissue programmed to form precursor joints between bones.

The flattened hand plate is defined during the fourth week, as the mesenchymal blastemas for the shoulder girdle, arm, and forearm regions are distinguished. The central carpal region of the hand plate appears during the fifth week surrounded by a thickened crescentic flange that will form the digital rays.

Atlas of Developmental Field Anomalies of the Human Skeleton: A Paleopathology Perspective, First Edition. Ethne Barnes.
© 2012 Wiley-Blackwell. Published 2012 by John Wiley & Sons, Inc.

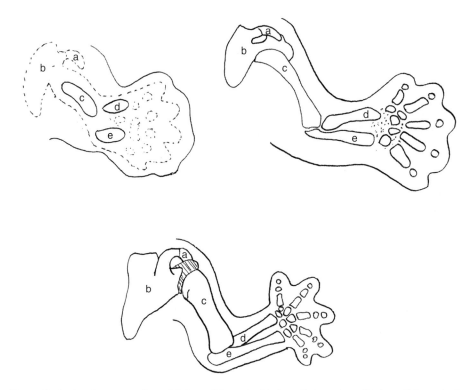

F I G U R E E-1.0. Upper limb development: from 38 days (11 mm), 41 days (14 mm) to 43 days (16 mm): dotted lines show continuous mesenchymal condensations, solid lines outline limb formation and precursor membranous bone formations—(a) clavicle, (b) scapula, (c) humerus, (d) radius, (e) ulna.

The margin of the hand plate becomes deeply notched with grooves forming between the digital rays in the sixth week. The limb rotates from the coronal plane toward the parasagittal plane. The elbows and wrists take a flexed position during the seventh week as swellings of the distal ends of the finger digits, separated by cell death of the interdigital tissue, form tactile pads. By the end of the eighth week, all bone segments are well defined in hyaline cartilage, with the single exception of the clavicles that begin to ossify from the membranous tissue in the sixth week.

The paired limb buds generally follow a synchronized timing of developmental events that creates a symmetrical pair of limbs, but it is not unusual for timing to be off, leaving one limb (usually the right) somewhat shorter than the other. Asymmetry often goes unnoticed and is insignificant. However, occasionally, one limb can be much shorter than the other, or just one segment of the limb can be affected, but usually does not affect the functioning of the limb. Unilateral anomalies usually develop on the right side. Complete or incomplete duplication of limb bones can occur with the duplicate most often a remnant of the other, but this phenomenon is very rare except in the digits. Complete absence of a developing limb (amelia) or part of the upper limb (meromelia—congenital amputation) is very rare.

SHOULDER GIRDLE SEGMENT

E-1. CLAVICLE DEVELOPMENT

The primordial clavicle arises and separates from the proximal end of the limb bud's continuous mesenchymal condensation. Two zones of membranous tissue develop within the mesenchymal condensation for the clavicle—the lateral or acromial zone and the medial or sternal zone—expanding toward each other, forming a bridge between the two in the middle zone (Fig. E-1.1A). Ossification begins directly from the two end zones. The two expanding ossifications cross the middle zone bridge and coalesce into the final bony format, while the mesenchymal ends turn into primitive hyaline cartilage that later ossify. Clavicle forms can be quite variable within the expected range of development. Anomalies are rare and usually affect the central portion between the acromial and sternal ends.

CLAVICLE ANOMALIES

E-1.1. Clavicle Hypoplasia/Aplasia

Unilateral or bilateral, this is often associated with cleidocranial dysostosis, a syndrome primarily affecting the

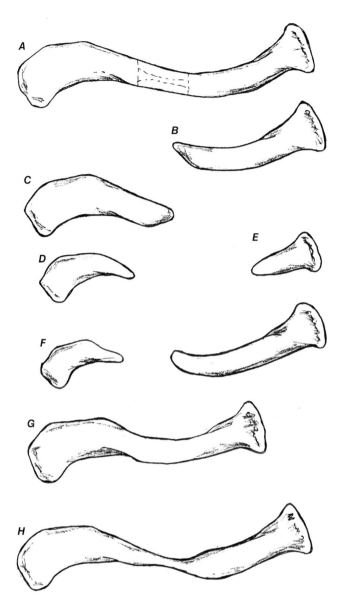

FIGURE E-1.1. Clavicle hypoplasia/aplasia: (right side) (*A*) normal clavicle—dotted lines represent the middle section of the original mesenchymal bridge between the acromial and sternal sections; (*B*) aplasia of the acromial end; (*C*) aplasia of the sternal end; (*D,E*) hypoplasia of the acromial and sternal ends; (*F*) unilateral hypoplasia affecting the acromial end; (*G*) short clavicle from the absent middle section; (*H*) middle section hypoplasia.

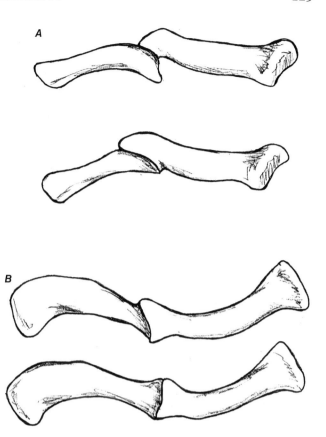

FIGURE E-1.2. Bifurcated clavicle (congenital pseudoarthrosis): (right side) (*A*) anterior view; (*B*) superior view—two divided clavicles showing the sternal ends overriding the acromial end with the pseudo joint.

membranous bone development of the skull and clavicle. Aplasia can be partial with either the bony acromial or sternal sections missing. The entire clavicle can appear unusually short if the center portion does not develop, or the middle portion can appear unusually thin from lack of expansion of the primordial bridge between the two sections. Hypoplasia of one or both the acromial and sternal sections can prevent the two from joining together, often leaving them connected by a fibrous band instead of bone (Fig. E-1.1).

E-1.2. Bifurcated Clavicle (Congenital Pseudoarthrosis)

Failure of the two ossified ends to connect in the middle section results in a clavicle consisting of two separate bones (Fig. E-1.2). The adjacent medial ends, covered by cartilage, form an articulation with a fibrous joint or a true cartilaginous joint. Usually, the longer sternal portion overrides the acromial portion (Manashil and Laufer 1979), not to be confused with a healed nonunion fracture of the clavicle.

E-1.3. Clavicle Duplication

This is the complete or incomplete partition of the membranous clavicle, a very rare phenomenon. It is either a separated remnant duplicate or incomplete separation dividing the acromial or sternal end (Fig. E-1.3). Separated remnants often split off from the acromion end and lie beneath the primary clavicle (Fig. E-1.4). Usually asymptomatic, the separated remnant may articulate with the primary clavicle, the coracoid process, or the humerus (Golthamer 1957; Twigg and Rosenbaum 1981).

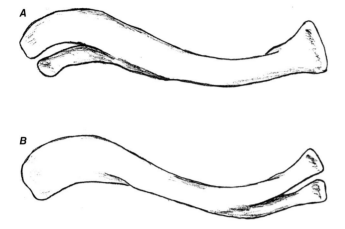

FIGURE E-1.3. Incomplete duplication of the clavicle: (right) (*A*) acromial end; (*B*) sternal end.

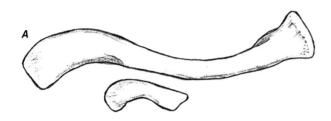

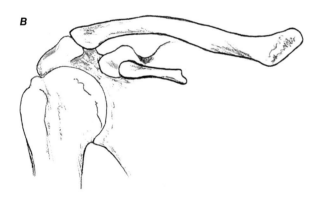

FIGURE E-1.4. Complete duplication of the clavicle: (right) secondary remnant acromial end (*A*) superior view; (*B*) ventral view showing the remnant articulating with the coracoid process.

E-2. SCAPULA DEVELOPMENT

This takes shape while separating from the distal portion of the shoulder segment evolving from the proximal portion of the continuous mesenchymal core of the upper limb. By the end of the fifth week, the shape of the body plate, the coracoid process, and the acromion are defined in condensed mensenchyme. By the eighth week, the contour of the mature scapula is set, and endochondral ossification begins within the body plate. At birth, the glenoid cavity, coracoid process, acromion, vertebral border, and inferior angle remain cartilaginous, programmed to ossify and fuse with the scapular body at specific times of growth and development (Fig. E-2.0).

By the end of the first year, the coracoid ossification center appears near its tip, while a separate ossification forms the root of the coracoid along with the upper one-third of the glenoid fossa rim (the subcoracoid) by 10–11 years. The bony coracoid unites with the body around 15 years, and the subcoracoid fuses with the body around 16–18 years. The lower two-thirds of the cartilaginous glenoid rim transforms into a bony epiphysis with the final fusion to the body between 20 and 25 years. The base of the acromion rises from an extension of the scapular spine, and by 15 years, several ossifying nuclei appear in the cartilaginous acromion, consolidating into three separate ossification centers, with the one nearest the base (meta-acromion) fusing to it immediately. The other two (meso-acromion and pre-acromion) quickly merge together into one bone that begins to unite with the rest of the acromion by 17–18 years, completed by around 20 years. The cartilage strip along the vertebral border and the inferior angle cartilage become bony epiphyses around 15 years, fusing to the body plate by 20 years.

SCAPULA ANOMALIES

E-2.1. Scapular Secondary Ossicles

These occur with failure of secondary ossifications to unite with the scapula, resulting in a separated scapular ossicle (Fig. E-2.1.1). The most common failure produces os acromion as the acromion formed by the distal two ossification centers fails to unite with the acromial base (Fig. E-2.1.2). The separated ossicle may have a true synovial joint or fibrocartilaginous connection to its base.

A separated coracoid can also occur but is very rare, with failure of the coracoid to unite with its root but remaining connected in a similar manner as the os acromion (Kim et al. 1998). Occasionally, the inferior angle bony epiphysis fails to fuse with the scapular body, united to it by fibrocartilage.

E-2.2. Scapula Secondary Ossification Hypoplasia/Aplasia

The pre-acromion at the tip of the acromion, or the combined pre-acromion–meso-acromion, may fail to ossify, leaving only the acromion base (Kim and Min

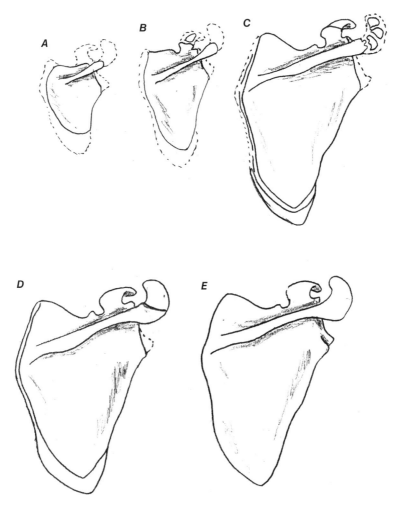

F I G U R E E-2.0. Scapula secondary ossifications: (dorsal right) broken lines represent unossified cartilage—(*A*) newborn; (*B*) ca. 3 years; (*C*) ca. 14 years; (*D*) ca. 15 years; (*E*) ca. 20 years.

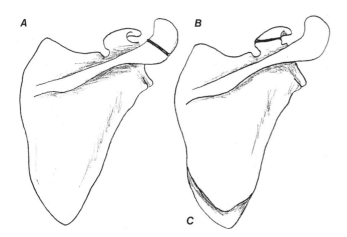

F I G U R E E-2.1.1. Scapula secondary ossicles: (dorsal right) failure to unite with the scapula—(*A*) os acromion; (*B*) os coracoid; (*C*) inferior angle ossicle.

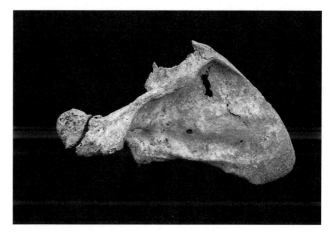

F I G U R E E-2.1.2. Scapula os acromion: left side, adult male, Ottoman Corinth, Greece.

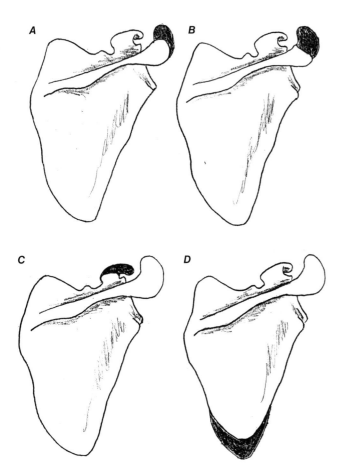

FIGURE E-2.2. Scapula secondary ossification hypoplasia/aplasia: (right dorsal) dark areas represent missing parts—(A) pre-acromion (tip) aplasia; (B) complete acromion aplasia; (C) coracoid tip aplasia; (D) inferior angle aplasia.

1994). Very rarely, the tip of the coracoid fails to ossify, leaving only the coracoid root in place. Occasionally, the inferior angle fails to ossify, leaving a short distal portion of the scapula (Fig. E-2.2).

E-2.3. Scapula Glenoid Neck Hypoplasia

This occurs when the bony epiphysis for the lower two-thirds of the crescent-shaped glenoid rim does not form (Fig. E-2.3), leaving the glenoid cavity more in appearance with the juvenile flattened shape. Normally, the adult glenoid cavity is slightly hollowed with a raised lower portion for attachment of the glenoid labium that deepens the cavity for the humerus head. This shortens the glenoid neck, and often, the glenoid cavity appears uneven as well as flattened. Usually bilateral and familial, it is often asymptomatic but can lead to joint stiffness, pain, and chronic joint instability (Brailsford 1948; Currarino et al. 1998).

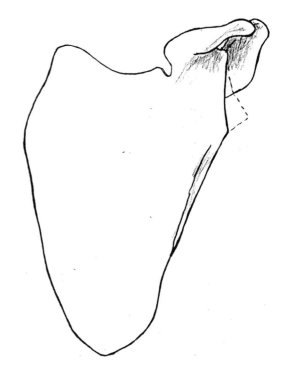

FIGURE E-2.3. Scapula glenoid neck hypoplasia: (costal left) broken lines represent the missing lower glenoid rim.

E-2.4. Scapular Aperture

This appears as a bony opening from incomplete primary ossification of a portion of the scapular body. It can be oval (Fig. E-2.4) or elongated in shape. The aperture is filled in with unossified fibrocartilaginous tissue (Cigtay and Mascatello 1979).

E-2.5. Sprengel's Deformity of the Scapula

This usually presents as a unilaterally elevated, hypoplastic scapula, with rotated axillary and vertebral borders. The embryonic scapula normally begins development opposite the primordial lower four cervical and upper thoracic vertebral segments. As the embryo grows, the scapula progressively descends below the lower cervical and first thoracic vertebral segments, beginning in the ninth week until it reaches its designated space by the third fetal month. Failure of the developing scapula to descend leaves it at the level of its origin. The upper part of the affected scapula often bends forward and hooks over the clavicle. About one-third of these defects have either a cartilaginous, fibrous, or osseous connection (omovertebral bone) with the transverse processes of associated cervical vertebrae (Fig. E-2.5). The affected shoulder has limited range of motion, especially abduction (Carson et al. 1981).

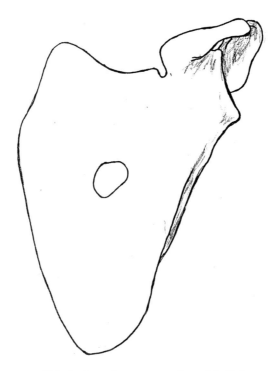

FIGURE E-2.4. Scapula aperture: (costal left) incomplete primary ossification, shape can vary from oval to elongated.

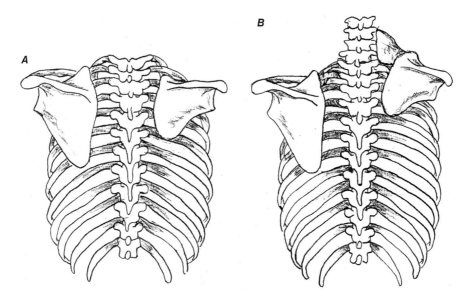

FIGURE E-2.5. Sprengel's deformity of the scapula: (*A*) affected right side without the accessory bone; (*B*) affected right side with the accessory omovertebral bone.

E-2.6. Scapular Coracoid–Clavicular Bony Bridge

The developing conoid and trapezoid ligaments connecting the superior border of the scapular coracoid to the inferior border of the clavicle are replaced by a band of preosseous cartilage that ossifies with bone maturity (Brailsford 1948) (Fig. E-2.6).

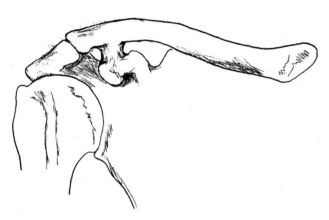

FIGURE E-2.6. Scapular coracoid–clavicular bony bridge: (right side) replacing conoid and trapezoid ligaments.

ARM SEGMENT

E-3. HUMERUS DEVELOPMENT

Primary ossification of the diaphysis begins during the eighth embryonic week, while the epiphyseal ends remain cartilaginous until after birth. Seven secondary ossification centers take form within the cartilaginous epiphyses at specific times of development, beginning with the proximal humeral head during the first year (Fig. E-3.0). Ossification of the greater tubercle begins in the third year, followed by ossification of the lesser tubercle during the fifth year. By the sixth year, all three ossifications combine into a single proximal epiphysis for the humeral head that eventually completely unites to the diaphysis around 20 years. A cartilaginous strip forms the distal epiphysis of the infant humerus from which four separate ossification centers take form. The first ossification center appears around 1.5 years—the capitulum and the radial border of the trochlea. The main portion of the trochlea begins to ossify around 12 years, followed by the lateral epicondyle at 13–14 years. All three unite into one distal epiphysis between 14 and 16 years, with final union with the diaphysis by 17 years. Ossification of the medial epicondyle begins around 5 years and remains separated from the other

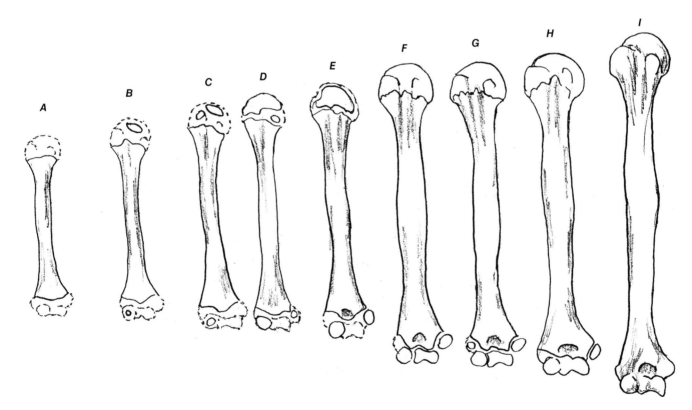

FIGURE E-3.0. Humerus development: (broken lines represent cartilage) (*A*) newborn; (*B*) 1–1.5 years; (*C*) third year; (*D*) fifth year; (*E*) sixth year; (*F*) twelfth year; (*G*) thirteenth year; (*H*) 14–16 years; (*I*) 18 years.

distal ossifications by a nonarticular area, with final union to the diaphysis around 18 years.

Developmental anomalies of the humerus are generally limited to minor variants that involve the complex secondary ossifications, with only very rare expressions of severe defects.

HUMERUS ANOMALIES

E-3.1. Phocomelia

Very rarely, the arm segment fails to develop, proximal phocomelia, or the forearm segment fails to develop, distal phocomelia, or both segments fail to develop, complete phocomelia, but usually, a complete or incomplete hand plate manages to form under the direction of the AER. The developing hand attaches to the remaining limb segment, or if both segments are absent, the hand attaches to the shoulder girdle or trunk in some fashion (Swanson et al. 1983).

Proximal Phocomelia (Agenesis of the Humerus)

Failed development of the humerus segment within the upper limb mesenchymal core disrupts developmental signals to the remaining forearm segment. The signaling disorders are variable, creating a range of malformed remnants of the radius and ulna that usually fuse together into a single, small irregularly shaped bone. Carpal development is disrupted and malformed, while digital development may be complete or incomplete, ranging in number from a few to all five digits (Figs. E-3.1.1 and E-3.1.2). The scapula is usually malformed and hypoplastic as the defective limb attaches to the trunk by way of a fibrous band, sometimes attaching to the clavicle.

Distal Phocomelia (Agenesis of the Forearm)

Absence of the forearm section primarily disrupts the development of the distal end of the humerus (Fig. E-3.1.3). The humerus itself is usually hypoplastic, as well as the scapula. The distal end takes on a distorted, bifurcated appearance, as if under the influence of the caudal–cranial axis that directs the development of the postaxial (ulnar) ray and preaxial (radial) ray. Carpal development is abnormal, and the number of digits is reduced or more likely to be missing than with proximal phocomelia (Denninger 1931).

E-3.2. Proximal Humeral Head Disturbance

Complete absence is extremely rare, but hypoplasia is known to occur (Fig. E-3.2), especially with scapular glenoid neck hypoplasia (Currarino et al. 1998).

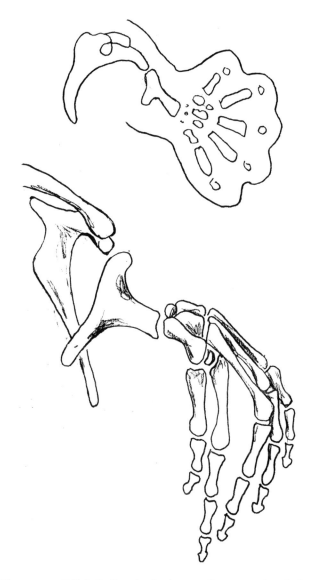

FIGURE E-3.1.1. Proximal phocomelia: agenesis of the humerus with malformed and fused bones of the forearm, malformed carpals with fully developed hand bones (drawn from Kelikian 1974).

E-3.3. Distal Humerus Disturbances

Supracondylar Process

This occurs when a small bony spur forms about 5 cm proximal to the distal medial epicondyle (Figs. E-3.3.1A and E-3.3.2). It is often joined by a fibrous band connected to the epicondyle that may also ossify. The two form a ring or canal for passage of the medial nerve and brachial artery (Barnard and McCoy 1946).

Septal Aperture

This appears with incomplete ossification of the thin bony wall between the distal olecranon and coronoid fossae above the trochlea (Figs. E-3.3.1B and E-3.3.3).

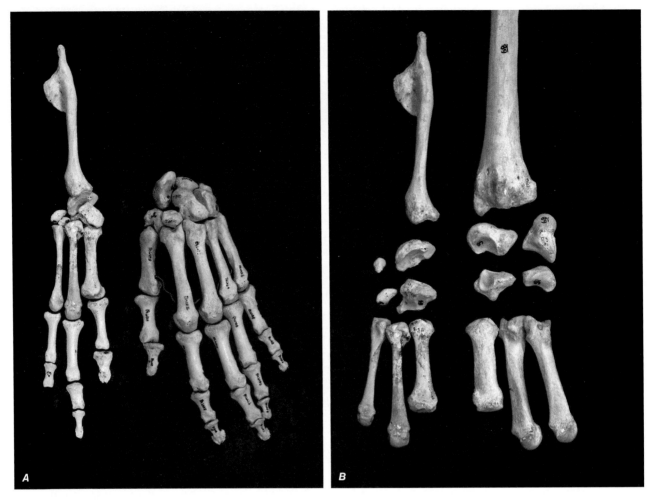

FIGURE E-3.1.2. Proximal phocomelia: (*A*) unilateral right with malformed radius and fused ulnar piece, four malformed carpals—scaphoid, trapezium, trapezoid, and capitate, with intact first three digits compared with the normal left hand; (*B*) separated malformed bones compared with similar bones from normal left side, older male adult (courtesy of Maxwell Museum anatomical collection).

The bony opening is filled with fibrocartilaginous tissue that failed to ossify.

Nonunion of Distal Secondary Ossifications
Most often, the medial epicondyle fails to unite with the diaphysis (Fig. E-3.3.1C). Very rarely are the other secondary ossifications contained within the joint capsule disunited.

Aplasia of Distal Secondary Ossifications
These usually involve the medial epicondyle that lies outside the distal joint capsule (Figs. E-3.3.1D and E-3.3.4), thus not affecting articulation with the bones of the forearm. Very rarely, the articulating capitulum, trochlea, or lateral epicondyle that is contained within the joint capsule fails to ossify, disrupting and displacing the proximal end of the associated forearm bone. Absence of the lateral epicondyle alone can displace the head of the radius.

E-3.4. Elbow Patella Cubiti

This is an accessory ossicle much like a true sesamoid bone, occurring in the triceps tendon at the lower dorsal end of the humerus, just above its attachment to the ulnar olecranon above the elbow joint (Fig. E-3.4). Articular cartilage has been identified underlying this accessory bone (Levine 1950; Sachs and Degenshein 1948).

FOREARM AND HAND SEGMENTS

PARAXIAL DEVELOPMENT

The development of the radius and ulna of the forearm, along with the proximal carpals within the forearm segment, is closely aligned to the development of the hand segment that includes the distal carpals. As the forearm segment evolves along the proximal–distal axis, it is greatly influenced by genetic signals for the

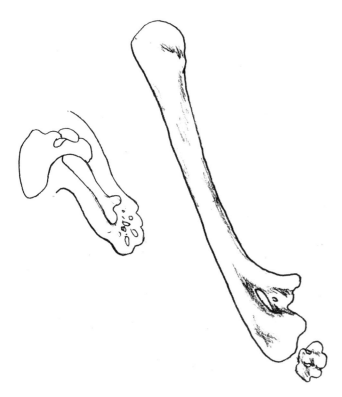

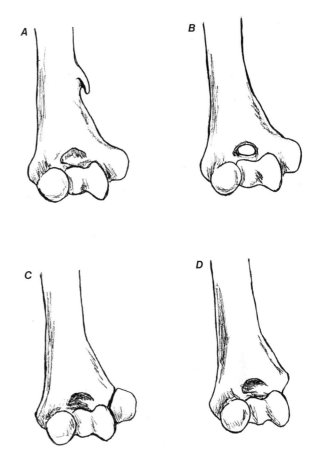

FIGURE E-3.1.3. Distal phocomelia: agenesis of the forearm with hypoplasia of the humerus, malformed distal humeral end with fused malformed carpals, absent hand (drawn from Denninger 1931).

FIGURE E-3.3.1. Distal humerus disturbances: (*A*) supracondylar process; (*B*) septal aperture; (*C*) medial epicondyle ossicle; (*D*) medial epicondyle aplasia.

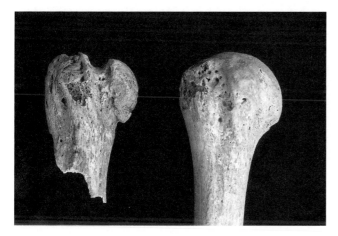

FIGURE E-3.2. Proximal humerus head hypoplasia: (right) compared with normal, adult male, Frankish Corinth, Greece.

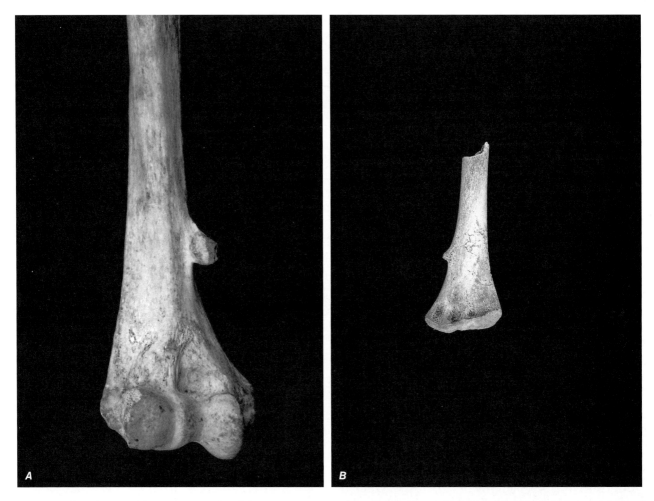

FIGURE E-3.3.2. Distal humerus supracondylar process: (*A*) (right), adult male (NMNH 271763); (*B*) (left) infant 24 months, Frankish Corinth, Greece.

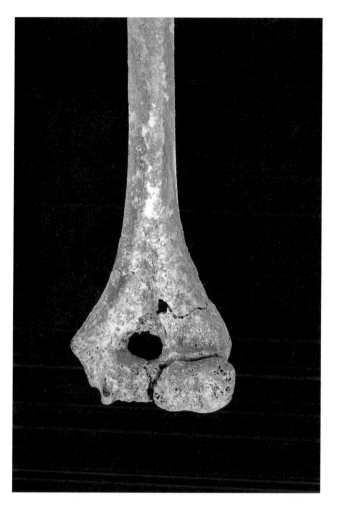

FIGURE E-3.3.3. Distal humerus septal aperture: (right) preadolescent male 12–13 years, La Playa, NW Mexico.

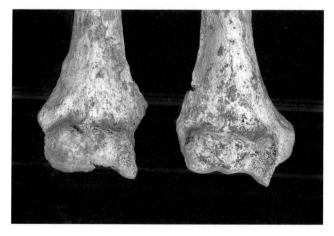

FIGURE E-3.3.4. Distal humerus medial epicondyle aplasia: (right) compared with normal, adult, Roman Corinth, Greece.

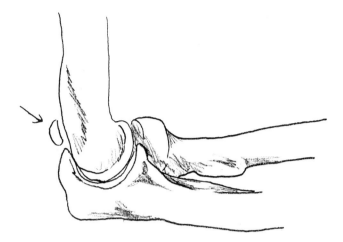

FIGURE E-3.4. Elbow patella cubiti: arrow points to the accessory ossicle above the elbow within the triceps tendon, dorsal side of the distal humerus, above the ulnar olecranon.

anterior–posterior axis coming from the ZPA adjacent to the AER on the dorsal side of the developing limb bud. This axis directs the development of the preaxial (radial) and postaxial (ulnar) rays of the forearm and hand segment, interacting with the genetic signals coming from the proximal–distal axis, as well as genetic signals affecting polarity coming from the dorsoventral axis. This complex of genetic signaling interactions between the axes is necessary for the development and positioning of the separate forearm bones, the distal and proximal carpals, and the metacarpals and phalanges of the digits. This very complexity invites aberrations, especially in the developing hand plate. Various classifications defining various types of hand and forearm anomalies can be found throughout the medical literature.

The preaxial (radial) ray primarily includes the radius, proximal scaphoid carpal, distal trapezium carpal, first digital ray (thumb), and secondarily the distal trapezoid carpal with the second digital ray. The postaxial (ulnar) ray includes the ulna, proximal triquetrum and pisiform, and the distal hamate carpal with the fifth and fourth digital rays. The third digital ray, along with the central distal capitate carpal and proximal lunate carpal, forms a central axis between the other two axes. The central axis can be influenced by the anomalous development of either one of the other two, but often tends to lean toward postaxial (ulnar) influence more than preaxial (radial) influence (Fig. E-4.0).

E-4. RADIUS AND ULNA DEVELOPMENT

At birth, the proximal head of the radius, the summit of the proximal ulna, and the distal ends remain in

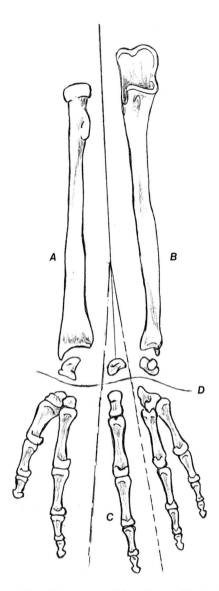

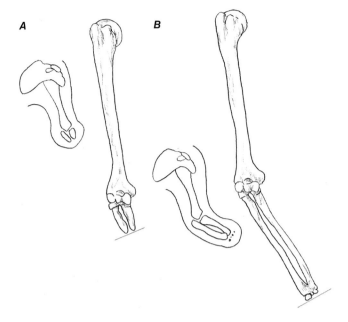

FIGURE E-4.1. Forearm meromelia (congential amputation): transverse limb deficiency (*A*) below the elbow; (*B*) distal end of the forearm segment.

FIGURE E-4.0. Forearm and hand paraxial development: (a) preaxial (radial) side—radius, proximal scaphoid, distal trapezium, and digital thumb ray, distal trapezoid with the second digital ray; (b) postaxial (ulnar) side—ulna, proximal triquetrum and pisiform, distal hamate, the fifth and fourth digital rays; (c) central axis—proximal lunate, distal capitate, and the third digital ray; (d) transverse line dividing the forearm segment and hand segment.

cartilage. The ossification center for the radial distal epiphysis appears at the end of the first year, followed by the ossification center for the proximal epiphysis from 4 to 5 years. Sometimes, a thin epiphysis with ossification forms on the radial tubercle around 14–15 years and quickly fuses with the shaft. The radial proximal end unites with the shaft between 14 and 17 years, followed by the distal end union between 17 and 19 years. The distal ulnar epiphysis begins to ossify between 5 and 6 years. As the bulk of the ulnar olecranon of

the proximal end develops as an extension of the shaft, ossification of the summit epiphysis occurs between 9 and 11 years, and unites with the olecranon between 14 and 16 years. The ulnar dorsal epiphysis unites with the shaft between 17 and 18 years. Rarely do the bony epiphyses fail to develop or remain separated from the mature shaft.

RADIUS AND ULNA ANOMALIES

E-4.1. Forearm Meromelia (Congenital Amputation)

This is described as transverse limb deficiency of the forearm limb segment where growth and development below a certain point are stopped (Fig. E-4.1). Very rarely, the arm or hand segments can also be disrupted in a similar manner. Most often, the disruption occurs below the elbow. The radius and ulna appear similar to amputated stumps, with the ends often united (Gladykowska-Rzeczycka and Mazurek 2009). The line of demarcation may also occur at the distal ends of the forearm (Swanson et al. 1983). With no further distal limb development, the hand plate fails to develop.

E-4.2. Forearm Paraxial Hemimelia

This is the agenesis (complete or partial) of the radial (preaxial) or ulnar (postaxial) side, but usually appears unilateral. A fibrocartilaginous band substitutes for the

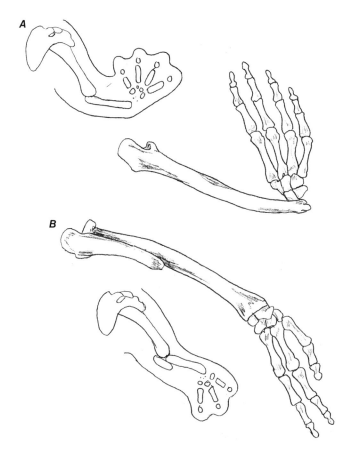

FIGURE E-4.2.0. Forearm paraxial hemimelia: (*A*) radial (preaxial) hemimelia—complete agenesis of the radius, scaphoid, trapezium, and digital thumb ray; (*B*) ulnar (postaxial) hemimelia—incomplete agenesis of the distal ulna, absent triquetrum, pisiform, hamate, and the fourth and fifth digital rays.

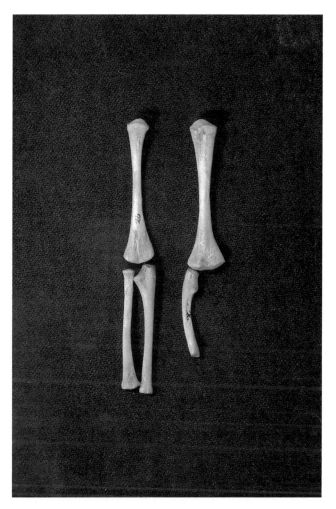

FIGURE E-4.2.1. Radial (preaxial) hemimelia: normal right side compared with the left side agenesis of the radius with a bowed, shortened ulna, newborn infant (courtesy of Maxwell Museum anatomical collection).

missing bone or missing bone segment, suggesting arrested development of the embryonic tissue precursor for the absent bone. Paraxial carpals and digital rays connected to the absent or partially absent forearm bone may be defective or fail to develop (Swanson et al. 1983) (Fig. E-4.2.0). The scapula is usually smaller, and the clavicle and humerus are shorter than normal.

Radial (Preaxial) Hemimelia

This is complete more often than partial (with partial absence, the distal portion is absent and the remaining bone is bowed). The ulna is thick and short, with a bowed distal end that dislocates dorsally and ulnarward (Fig. E-4.2.1). The scaphoid, trapezium, and digital thumb ray are absent. The second digital ray, trapezoid, and lunate may also be missing. With lack of support from the radius, the wrist becomes unstable and unable to support the hand, and it deviates to the radial side (Fig. E-4.2.0A) at an acute angle to the ulna (radial club hand).

Ulnar (Postaxial) Hemimelia

This is seen far less often than radial hemimelia and usually presents as partial agenesis of the distal portion of the ulna with bowing of the remaining bone portion (Fig. E-4.2.0B). The radius is also bowed and often short, but can be of normal length. The proximal radius and distal humerus may be conjoint with the forearm flexed at an angle (Figs. E-4.2.2 and E-4.2.3), or the radial head becomes displaced on the humerus. The fifth and fourth digital rays, triquetrum, pisiform, and hamate are absent, and the third digital ray with the lunate and capitate may also be affected. Sometimes, only the digital thumb and associated carpals are present (monodactyly). The hand deviates to the ulnar side at a slightly flexed angle (ulnar club hand) (Fig. E-4.2.0B).

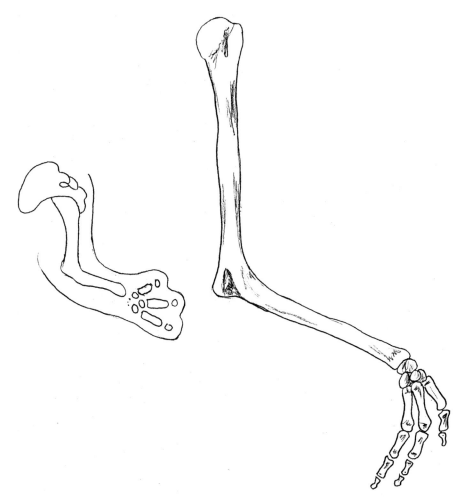

FIGURE E-4.2.2. Ulnar (postaxial) hemimelia: complete agenesis of the ulna, triquetrum, pisiform, hamate, and the fourth and fifth digital rays, with conjoined humerus and radius.

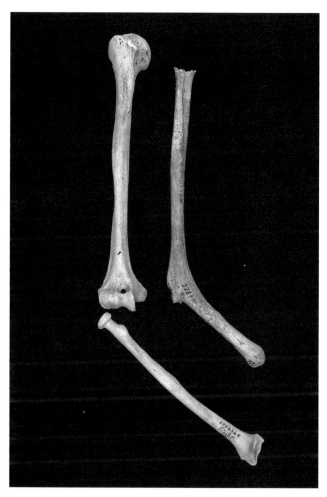

FIGURE E-4.2.3. Ulnar (postaxial) hemimelia: (right) complete agenesis of the ulna with the humerus conjoined to a shortened radius (NMNH 228400), Moundville, AL, compared with the normal right radius and humerus.

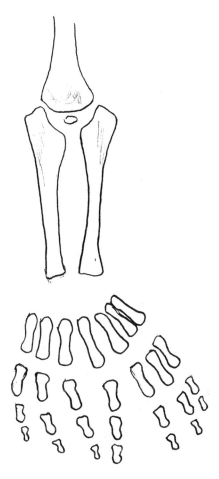

FIGURE E-4.3. Duplication (dimelia) of the ulnar (postaxial) ray: agenesis of the radius, first digital ray, mirror hand duplication remaining digital rays on each side of the central digital ray (infant).

E-4.3. Duplication (Dimelia) Forearm Ray

This occurs very rarely and usually involves the duplication of the ulnar ray while the radial ray does not develop. The ulnar ray duplicates itself with the two ulnae turned toward each other (Fig. E-4.3). The distal humerus lacks a capitulum, and instead has two poorly defined trochleas with doubled olecranon fossa. The postaxial (ulnar) digital rays and carpals are also duplicate, usually with a central digital ray in place with duplicates on each side, producing a "mirror hand" of seven or more digits, lacking preaxial (radial) digital rays and carpals (Kelikian 1974).

E-4.4. Madelung's Deformity

This does not become apparent until early adolescence. Growth of the medial border of the radius lags behind the growth of the other side of the radial border, bending the radius so that the distal end curves in a volar direction toward the ulna (Anderson and Carter 1995; Canci et al. 2002). The ulna appears longer than the shortened, curved radius, and dislocates backward, with the ulnar head forming a large knotty protrusion on the backside of the wrist. This anomaly creates a peaked V-shaped dome interface for the articulating proximal carpals instead of a convex interface (Fig. E-4.4). The lunate locates at the apex of this dome, sandwiched in between the forearm bones. The remaining carpals are wedged tightly together and pushed palmward toward the ulnar border (Kelikian 1974).

E-4.5. Radial–Ulnar Synostosis

This implies the union of the proximal ends of the two separate adjacent bones, but actually the proximal ends of these two bones were never separate. They remain conjoined with failure to part during embryonic development. The anterior–posterior axis directs the separation of the primordial mesenchymal core for the forearm

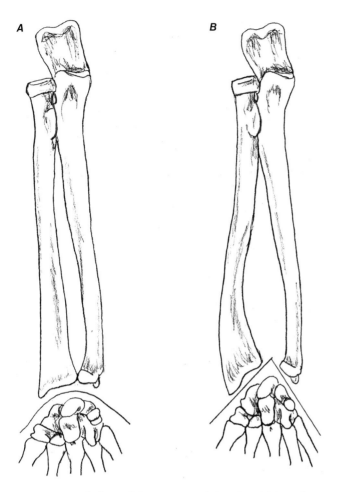

FIGURE E-4.4. Madelung's wrist deformity: (right palmer aspect) (*A*) normal right forearm with convex line representing normal wrist articulation; (*B*) deformed distal radius forcing an abnormal V-shaped wrist articulation with lunate at the apex.

into the ulnar and radial rays, beginning at the distal end and progressing toward the proximal end. As divergence nears completion, the distal end of the precursor radius crosses over the distal precursor ulna, while the proximal ends roll toward the ulnar coronoid process as an interzone forms between them for the formation of the radial–ulnar joint. If the interzone fails to develop, this leaves the two opposing proximal ends united by preosseous cartilaginous tissue (Fig. E-4.5), followed by ossification with a shared medullary canal. Often, there is aplasia or hypoplasia of the radial head with complete assimilation of the proximal radius into the ulna just below the coronoid process, extending about 4–8 cm. Bony union with the ulna can also take place with complete development of the radial head, as a bony bridge about 2–4 cm long connected to the proximal ulna unites them just below the radial head, causing it to grow away from the ulna. Bilateral, asymmetrical occurrence is more common than unilateral expression.

The forearm is fixed in pronation or semipronation, restricting supination, placing severe strain on the elbow and shoulder (Kelikian 1974).

E-4.6. Ulnar Styloid Os/Aplasia

Often, a separate ossification center forms for the styloid process that may or may not unite with the rest of the distal epiphysis, or the styloid does not ossify at all (Fig. E-4.6).

E-5. CARPUS DEVELOPMENT

Eight irregularly shaped small bones are separated by synovial joints, closely packed together, and held in place by interosseous ligaments within two transverse rows. The proximal carpals, from the radial to ulnar side—scaphoid, lunate, and triquetrum (with pisiform interfacing its palmer surface)—together form the uneven midcarpal joint with the distal carpals, from the radial to ulnar side—trapezium, trapezoid, capitate, and hamate. The proximal scaphoid and lunate interface with the radius, while the triquetrum interfaces with the radiocarpal disk at the distal end of the ulna, forming a single capsulated, radiocarpal joint (Fig. E-5.0). These two composite joint systems allow the wrist complex greater ranges of motion capable of withstanding imposed pressures from the hand. The proximal palmar pisiform, not part of the radiocarpal joint, behaves like a sesamoid bone contained within the tendon for the flexor carpi ulnaris, resting on the palmer surface of the triquetrum. The pisiform is the only carpal with a muscle attachment. Each distal carpal forms a separate synovial joint with each corresponding interfacing metacarpal.

The individual carpal bones vary from one another in shape as they facilitate multiple articulations with adjacent bones, except for the pisiform that has only one articulation. There is also size and shape variation within set patterns of development for each carpal. The pisiform can have variable positioning on the triquetrum, sometimes projecting beyond the ulnar side of the bone. During embryonic development, each carpal evolves from more than one mesenchymal nuclei before consolidating into single preosseous cartilaginous centers for each carpal. Sometimes, more than one preosseous center develops within one carpal, but usually, they unite as one before ossification is complete. Otherwise, a small accessory ossicle forms that may or may not fuse with the parent carpal.

Developmental anomalies of the carpals most often go undetected in immature human skeletal remains until the bones reach maturity. At birth, the carpals are cartilaginous. By 6 weeks, the first ossification center appears in the capitate, followed by the hamate at 3–4

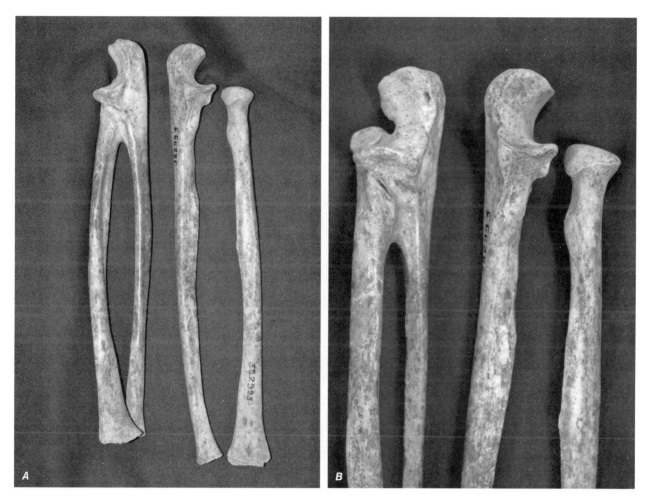

FIGURE E-4.5. Radial–ulnar synostosis: (*A*) right conjoint proximal ends with divergent distal radius crossing over the distal ulna, compared with normal left side, (*B*) close-up, adult (NMNH 382993), Mobridge, SD.

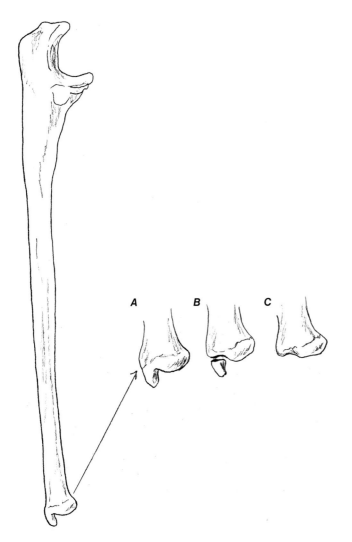

FIGURE E-4.6. Ulnar styloid os/aplasia: (*A*) normal distal ulna with the styloid process; (*B*) ulnar styloid os; (*C*) absent styloid process.

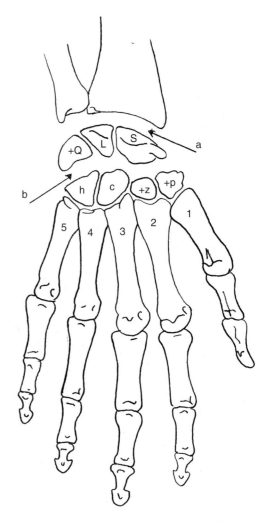

FIGURE E-5.0. Carpal major joints: (dorsal aspect right wrist) (a) radiocarpal joint; (b) midcarpal joints. S = scaphoid; L = lunate; tq = triquetrum; h = hamate; c = capitate; tz = trapezoid; tp = trapezium.

months, then the triquetrum between 7 and 24 months. Ossification of the lunate begins between 3 and 4 years, followed by the scaphoid, trapezium, and trapezoid before 6 years. The pisiform begins ossification between 8 and 12 years, with full maturity of all carpals between 14 and 18 years.

CARPAL ANOMALIES

E-5.1. Carpal Coalitions

This most often occurs just between two adjacent carpals, either bilateral or unilateral, as the interfacing interzone for the joint fails to appear during morphogenesis (Garn et al. 1976). Depending on the extent of the disruption within the interzone, the carpal union can be a complete (Fig. E-5.1.1) or incomplete osseous or nonosseous union. With nonosseous union, a fibrocartilaginous connection replaces the primordial synovial joint between the affected carpals, reflected by pitting lesions on interfacing carpal surfaces (Resnik et al. 1986) that can be confused with osteoarthritis. Coalitions usually involve two adjacent carpals in the same transverse row (Fig. E-5.1.2), very rarely cross over the midcarpal joint to the next row (Oner and de Vries 1994).

E-5.2. Atypical Carpal Coalitions

Midcarpal, carpal–metacarpal, radiocarpal coalitions can occur. Joint failure may be seen between the radius and either the lunate or the scaphoid, or both. Union

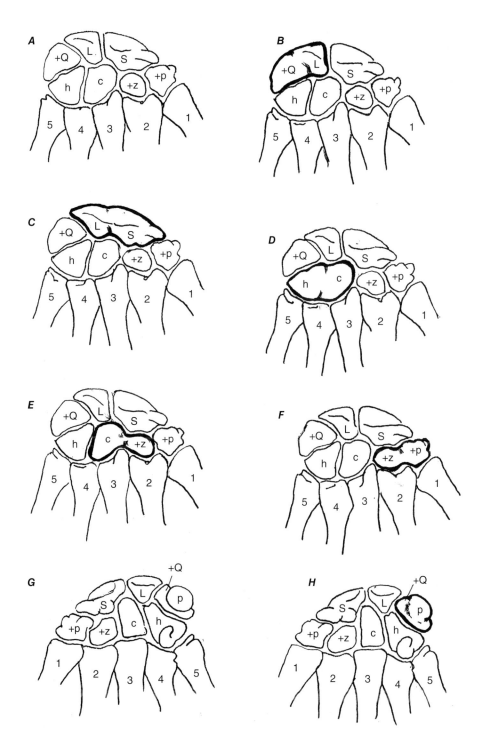

FIGURE E-5.1.1. Carpal coalitions: (right wrist) dorsal aspect—(*A*) normal; (*B*) lunate-triquetrum union; (*C*) lunate-scaphoid union; (*D*) hamate-captitate union; (*E*) capitate-trapezoid union; (*F*) trapezium-trapezoid union; palmar aspect—(*G*) normal; (*H*) trapezium-pisiform union. S = scaphoid; L = lunate; tq = triquetrum; h = hamate; c = capitate; tz = trapezoid; tp = trapezium; p = pisiform.

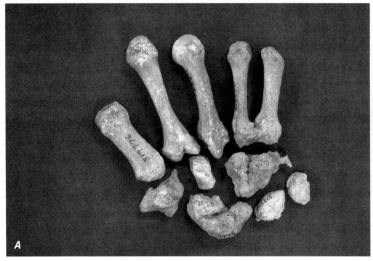
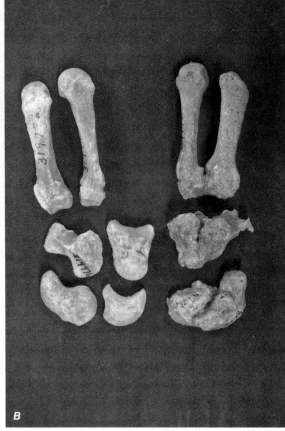

F IGURE E-5.1.2. Carpal coalitions double: (*A*) scaphoid–lunate, capitate–hamate with conjoined fourth to fifth metacarpals, unilateral right wrist; (*B*) coalitions compared with the normal left carpals, fourth, fifth metacarpals, adult female (NMNH 319776), England (note the deformed hamulus on the affected hamate).

between a distal carpal and its corresponding metacarpal can also occur. Coalition across the midcarpal joint between a proximal and distal carpal has been noted on rare occasions. The most unusual is the midcarpal union between the pisiform and hamulus of the hamate (Burnett 2011), which appears to replace the pisohamate ligament connecting the two entities. Any of these unions can be complete or incomplete, osseous (Fig. E-5.2.1) or fibrocartilaginous with pitting of the affected interfacing bony surfaces.

Massive Carpal Coalition
This is a very rare absence of joint development involving all or nearly all carpals, except the pisiform. This often involves the union of most if not all carpal–metacarpal joints as well, except for the first metacarpal of the thumb (Tuncay et al. 2001) (Figs. E-5.2.2–E-5.2.6).

E-5.3. Carpals Bipartite and Separated Marginal Carpal Elements

Very rarely, a carpal is divided by the development of two separate preossification centers that fail to unite

during morphogenesis. Fibrocartilage, instead of a true synovial joint, usually fills the gap between them (Boyd 1933; Kelikian 1974). Separation can be complete or incomplete (Figs. E-5.3.1 and E-5.3.2). The scaphoid is affected more than any other carpal, followed in order by the triquetrum, pisiform, trapezium, trapezoid, and capitate. Shapes of the two parts may appear distorted, making it difficult to identify bifurcated carpals in disarticulated human skeletal remains (Fig. E-5.3.3). A separate ossification center for the hamulus of the hamate can develop, leaving it as a separate ossicle. The tuberosity of the scaphoid may also develop from a separate ossification center and remain separated from the rest of the bone.

E-5.4. Carpal Hypoplasia/ Aplasia/Hyperplasia

When the entire carpals are affected, it usually occurs with segmental disturbances (see Fig. E-3.1.2), and isolated carpals so affected are very rare. However, Kelikian (1974) identified a hypoplastic scaphoid on one wrist of an adult male, with aplasia of the scaphoid on the

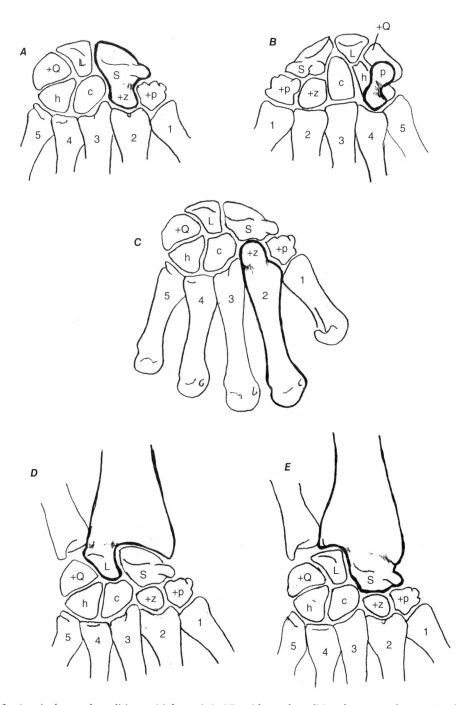

FIGURE E-5.2.1. Atypical carpal coalitions: (right wrist) (*A*) midcarpal coalition between the proximal scaphoid and distal trapezoid; (*B*) midcarpal coalition between proximal pisiform and hamate of the distal hamulus; (*C*) carpal–metacarpal coalition between the trapezoid and second metacarpal; (*D*) radiocarpal coalition between the lunate and radius; (*E*) radiocarpal coalition between the scaphoid and radius. S = scaphoid; L = lunate; tq = triquetrum; h = hamate; c = capitate; tz = trapezoid; tp = trapezium; p = pisiform.

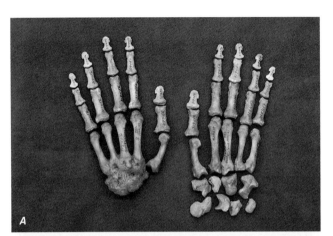

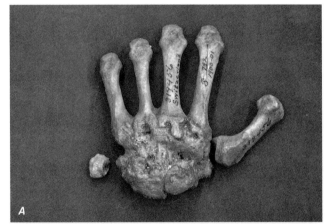

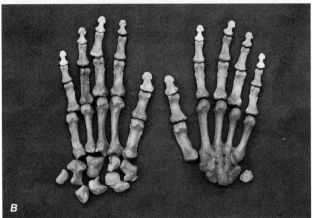

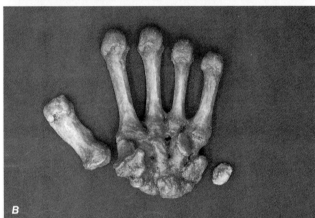

FIGURE E-5.2.2. Massive carpal coalition: (*A*) dorsal view, (*B*) palmar view—unilateral left hand with the normal right hand—all carpals except the pisiform and all metacarpals except MC1 united, adult male (NMNH 319456), Switzerland (identified by Kristen Pearlstein and Kathleen Adia).

FIGURE E-5.2.3. Massive carpal coalition: (*A*) dorsal view, (*B*) palmar view—close-up of the left wrist of Figure E-5.2.2, adult male (NMNH 319456), Switzerland.

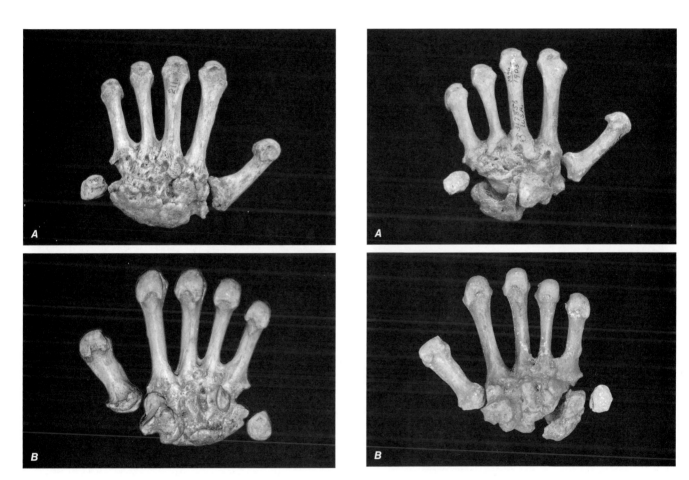

FIGURE E-5.2.4. Massive carpal coalition: (*A*) dorsal view, (*B*) palmar view—unilateral left hand, all carpals except the pisiform and all metacarpals except MC1 united, adult male (211) NMNH Terry collection (identified by Kristen Pearlstein and Kathleen Adia).

FIGURE E-5.2.5. Massive carpal coalition: (*A*) dorsal view, (*B*) palmar view—unilateral left hand, with lunate–triquetrum coalition separate from the coalition of all other carpals, separate pisiform, main massive coalition united with all metacarpals except MC1, adult male (NMNH 317856), USA (identified by Kristen Pearlstein and Kathleen Adia).

148   CHAPTER E UPPER LIMBS

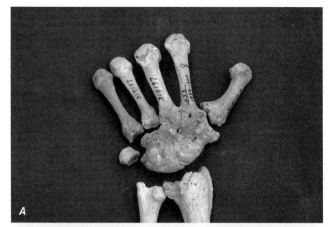

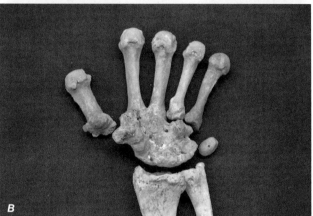

FIGURE E-5.2.6. Massive carpal coalition: (*A*) dorsal view, (*B*) palmar view—unilateral left hand, all carpals except the pisiform, united with the second and third metacarpals, adult female (NMNH 319197), Ireland (identified by Kristen Pearlstein and Kathleen Adia).

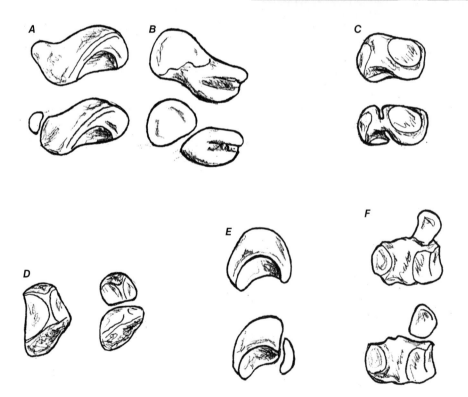

FIGURE E-5.3.1. Carpals bipartite and separated carpal elements: (right) (*A*) normal scaphoid (palmar aspect) and scaphoid with a separate tubercle; (*B*) normal scaphoid (dorsal aspect) and bifurcated scaphoid; (*C*) normal triquetrum (palmar aspect) and incomplete bifurcated triquetrum; (*D*) normal trapezoid and bifurcated trapezoid; (*E*) normal lunate and bifurcated lunate; (*F*) normal hamate and hamate with separated hamulus.

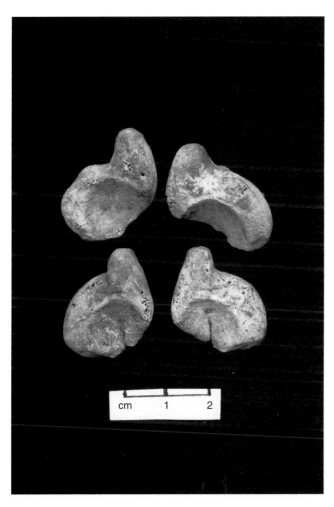

FIGURE E-5.3.2. Carpal incomplete bifurcation: bilateral scaphoids (bottom) compared with normal scaphoids (top), adult male, Frankish Corinth, Greece.

FIGURE E-5.3.3. Carpals bipartite: left side—normal right trapezoid (top) and bifurcated left trapezoid; right side—normal left lunate (top) and bifurcated right lunate with degenerative joint arthritis, adult female, Frankish Corinth, Greece.

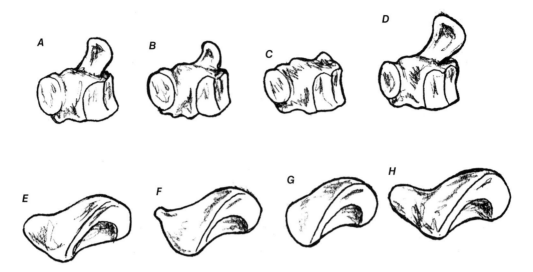

FIGURE E-5.4. Hypoplasia/aplasia/hyperplasia carpal elements: (*A*) hamate normal hamulus; (*B*) hamate hamulus hypoplasia; (*C*) hamate hamulus aplasia; (*D*) hamate hamulus hyperplasia; (*E*) scaphoid normal tubercle; (*F*) scaphoid tubercle hypoplasia; (*G*) scaphoid tubercle aplasia; (*H*) scaphoid tubercle hyperplasia.

other wrist. Hypoplasia/aplasia usually impacts only marginal elements of a developing carpal, when they arise from a separate ossification center. The hamulus of the hamate is the most affected, followed by the scaphoid tuberosity (Fig. E-5.4).

E-5.5. Os Metastyloideum

Occasionally, the third metacarpal styloid process develops from a separate preossification center that may or may not unite with the metacarpal. Failure of the union with the metacarpal places it in the carpus as an extra ossicle between the capitate and trapezoid, or it may unite with the capitate or the trapezoid (Figs. E-5.5.1 and E-5.5.2). Failure of ossification of the styloid process may also take place (Dwight 1907).

E-6. DIGITAL DEVELOPMENT

The metacarpals and phalanges form within the digital ray mesenchymal condensations as proximal–distal sequence segmentation within the hand plate takes place, primarily under the direction of the AER. Additional small mesenchymal condensations may also develop, and some unite with an evolving precursor digital bone segment. Some remain separate to develop into sesamoids within tendons attaching to the metacarpal–phalangeal joints, especially a pair of sesamoids at the plantar aspect of the first metacarpal joint. Primary shaft ossification begins in the eighth embryonic week from the cartilaginous tissue, while the distal

ends of the terminal phalanges ossify directly from the membranous tissue. Secondary ossification of the epiphyses between 2 and 4 years starts with the proximal epiphyseal base of the first metacarpal, followed by the distal epiphyseal heads of the remaining metacarpals, and the proximal epiphyseal bases of the phalanges. Occasionally, the first metacarpal may also have a separate epiphyseal head. Secondary ossifications begin to unite with their shafts during adolescence, completely fused by 18–19 years.

The very complexity of the digital rays invites a wide variety of anomalies, producing a range of mixed anomalous genetic patterns within the developing hand.

DIGITAL ANOMALIES

E-6.1. Brachydactyly

This is hypoplasia/aplasia of one or more segments within the same digital ray, shortening the affected finger. A wide range of patterns of various digital segmental disturbances appear along genetic lines, most often medically classified accordingly as type A (BDA) to type E (BDE) with subtypes for each (Bell 1951; Temtamy and McKusick 1978). Most important to paleopathology is to identify patterns of brachydactyly within a skeletal population rather than trying to fit them into a specific designated category (Figs. E-6.1.1 to E-6.1.3). The middle phalanges are affected the most with hypoplasia or aplasia. Phalangeal aplasia in ancient human skeletal remains can be difficult to decipher, not knowing if they were missed in recovery. Metacarpal

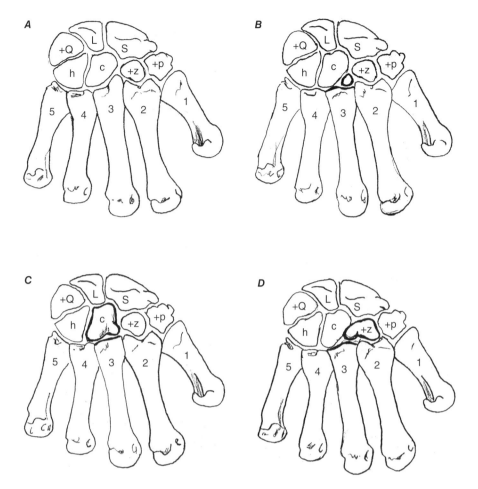

FIGURE E-5.5.1. Os metastyloideum: (right dorsal) (*A*) normal carpus, metacarpals; (*B*) separated third metacarpal styloid process ossicle; (*C*) separated styloid ossicle united with the capitate; (*D*) separated styloid ossicle united with the trapezoid. S = scaphoid; L = lunate; tq = triquetrum; h = hamate; c = capitate; tz = trapezoid; tp = trapezium.

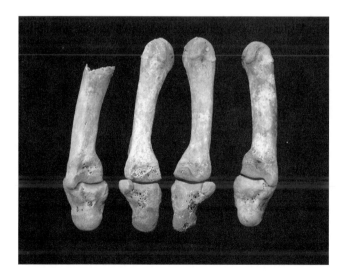

FIGURE E-5.5.2. Os metastyloideum: MC3 styloid separated ossicle united with the capitate bilateral (middle), adult female, compared with normal MC3s and capitates on the sides, adult female, Frankish Corinth, Greece.

hypoplasia leading to bone shortening (Fig. E-6.1.3) is the most readily detected form of brachydactyly (Case 1996).

Atypical Brachydactyly

This involves irregular shapes of shortened bones. This includes a hypoplastic triangular-shaped middle phalanx, known as the delta phalanx that usually affects the second finger, directing growth of the terminal phalanx toward the radial side (Jones 1964). Another atypical form primarily affects the middle phalanx of the fifth finger with a hypoplastic rhomboid shape that also directs the growth of the terminal phalanx to the radial side. A short, broad-based terminal phalanx, usually of the thumb known as "stub thumb," represents another atypical form. The terminal phalanx of the fifth finger can atypically develop Kirner's deformity with radial bowing of the shaft above its base that is skewed in development (David and Burwood 1972) (Fig. E-6.1.4).

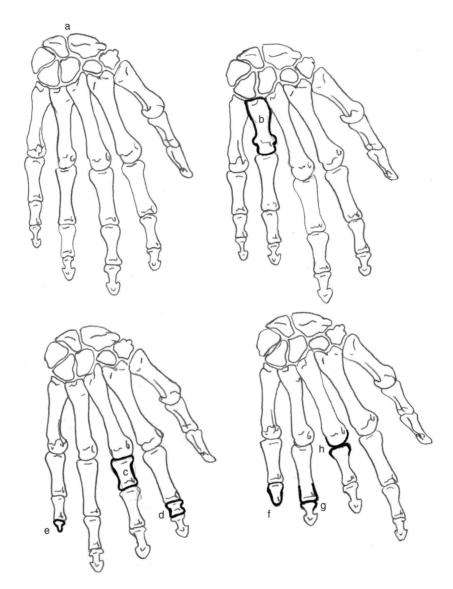

F I G U R E E-6.1.1. Brachydactyly—typical forms: (a) normal digits; (b) hypoplasia fourth metacarpal; (c) hypoplasia third proximal phalanx; (d) hypoplasia second middle phalanx; (e) hypoplasia fifth distal phalanx; (f) fifth distal phalanx aplasia; (g) fourth middle phalanx aplasia; (h) third proximal phalanx aplasia.

FIGURE E-6.1.2. Brachydactyly: right second distal phalanx hypoplasia, adult female, Frankish Corinth, Greece.

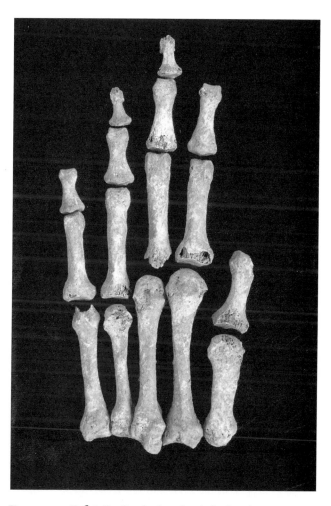

FIGURE E-6.1.3. Brachydactyly: left fourth metacarpal mild hypoplasia, adult male, La Playa, NW Mexico.

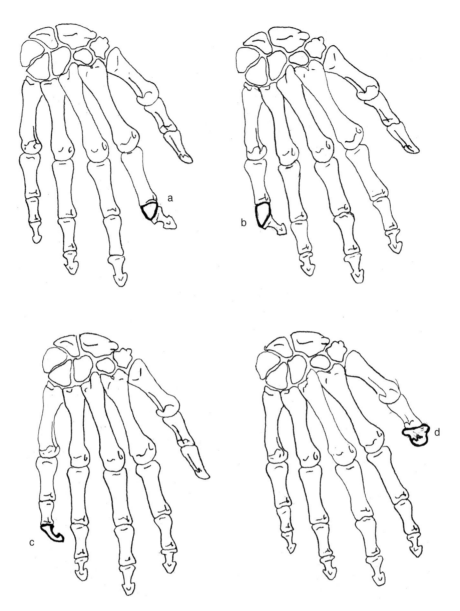

FIGURE E-6.1.4. Brachydactyly—atypical forms: (*A*) delta (triangular) second middle phalanx; (*B*) rhomboid fifth middle phalanx; (*C*) Kirner's deformity fifth distal phalanx; (*D*) "stub thumb" short, broad first distal phalanx.

E-6.2. Syndactyly Complex

This occurs when all or part of the adjacent digits fails to separate during morphogenesis. Most often, only the soft tissue of the digits fails to separate, resulting in webbed fingers joined by fascia. This is known as simple syndactyly and may involve more than two digits. Complex forms involve the bone tissue (Fig. E-6.2), easily recognized in human skeletal remains, varying from conjoined metacarpals (Dwight 1907) to conjoined distal phalanges, and often combined with other digital anomalies (Kelikian 1974; Temtamy and McKusick 1978).

E-6.3. Symphalangism

This is the failure of segmentation and interphalangeal joint formation between the phalanges within a digit. The affected digit can appear partially flexed as it does so while forming in membranous tissue during morphogenesis. More than one digit can be affected. Most often, the fifth and fourth digits are affected, followed by the third and second digits. Symphalangism of the thumb is very rare. Either the proximal (proximal symphalangism) or distal (distal symphalangism) interphalangeal joint is absent, or very rarely, both joints are not present (Fig. E-6.3). Proximal symphalangism, also known as

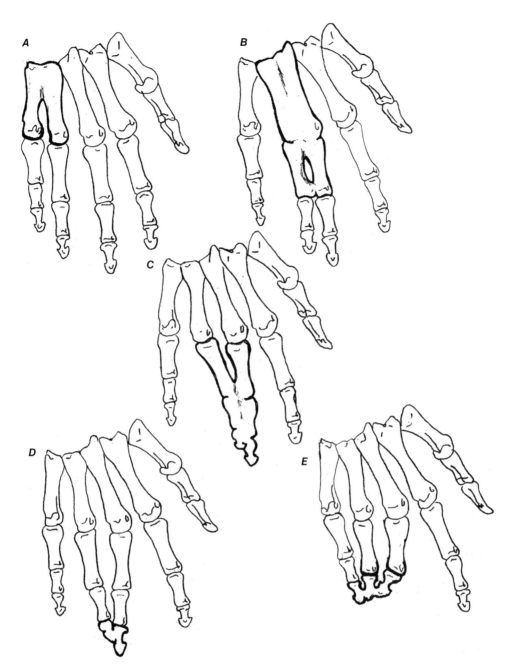

FIGURE E-6.2. Syndactyly complex: (*A*) partially conjoined fifth and fourth metacarpals; (*B*) conjoined third and fourth metacarpals and middle phalanges; (*C*) partially conjoined third and fourth proximal phalanges with completely conjoined middle and distal phalanges; (*D*) conjoined third and fourth distal phalanges; (*E*) conjoined third, fourth, and fifth distal phalanges with agenesis of the third and fourth middle phalanges.

Cushing's symphalangism, is more common in finger digits than distal symphalangism (Flatt and Wood 1975).

E-6.4. Triphalangeal Thumb

This is very rare. The thumb takes on characteristics of a finger digit with lengthening of the first metacarpal, a middle phalanx, and shortened distal phalanx (Fig. E-6.4). The epiphysis for the first metacarpal may be

distal or proximal (like finger digit metacarpals) or both. Two major types occur—opposing and nonopposing. The opposing type appears to receive mixed signals to be fingerlike with a rudimentary, often triangular (delta)-shaped extra ossicle for a middle phalanx, and nearly normal opposition with most thenar muscles developing. Signals for the development of the nonopposing type appear to replace thumb morphogenesis with the development of a finger digit (often looking like a

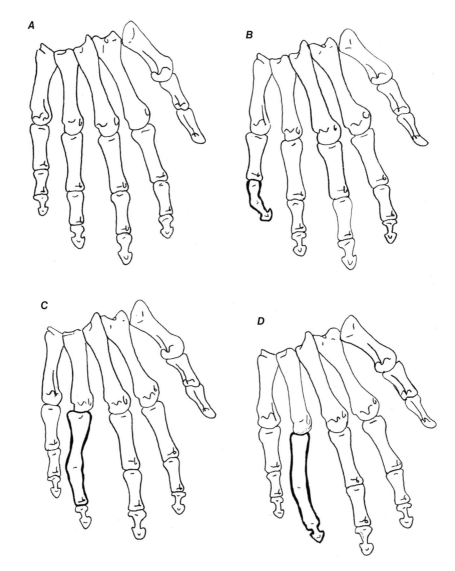

FIGURE E-6.3. Symphalangism: (*A*) normal right finger digits; (*B*) fifth digit distal interproximal joint union; (*C*) fourth digit proximal interproximal joint union; (*D*) fourth digit proximal and distal interphalangeal joint union.

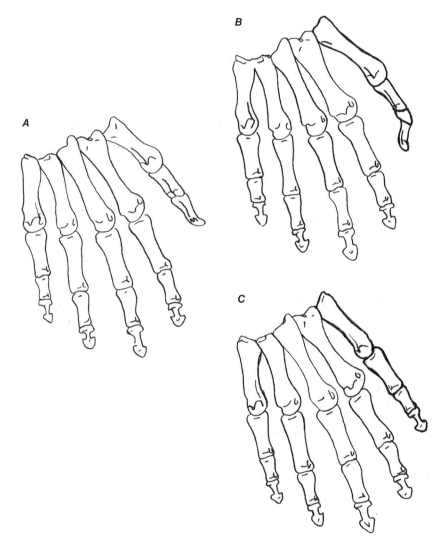

FIGURE E-6.4. Triphalangeal thumb: (*A*) normal; (*B*) triphalangeal opposing thumb; (*C*) triphalangeal nonopposing thumb (five-fingered hand).

mirror twin to the fifth digit) in place of the thumb, with absent thenar muscles. This type is known as the five-fingered hand (Brailsford 1948; Zguricas et al. 1997).

E-6.5. Ectrodactyly

This is the agenesis of all or part of one or more digital rays. While paraxial agenesis of the radial or ulnar rays (see Section E-4.2) includes agenesis of the carpals and finger digits on the affected sides, the absence of one or more digital rays can also occur with fully developed forearms. Many different patterns and combinations of various types of digital agenesis form when signals from

the AER for digital morphogenesis are disrupted. The metacarpals are not always involved, and very rarely are associated carpals included in the defect.

The most often recognized form of digital ray agenesis is the split (cleft) hand, usually missing all or parts of one to three middle digital rays, forming a central cleft that divides the remaining digits into ulnar and radial portions. Lobster claw hand is the best known expression of the many variants of split hand, usually with the entire central digital ray missing, including the metacarpal, and sometimes the capitate carpal, usually accompanied by syndactyly of the remaining two digits on each side of the cleft, forming the appearance of opposing claws (Figs. E-6.5.1 and E-6.5.2). This may or

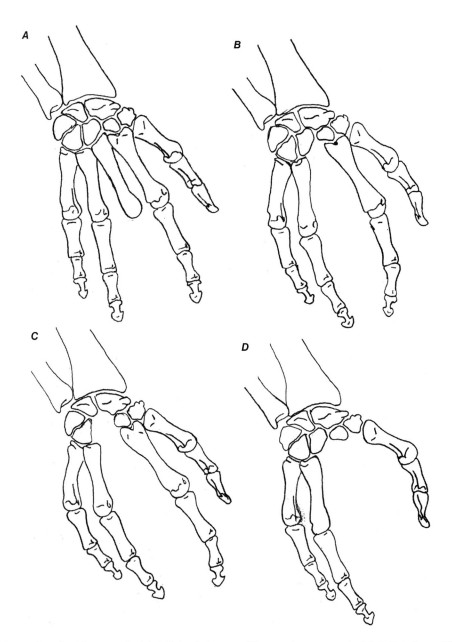

FIGURE E-6.5.1. Ectrodactyly: (*A*) agenesis third digit phalanges; (*B*) complete agenesis third digital ray; (*C*) complete agenesis third digital ray and capitate; (*D*) agenesis second and third digital rays.

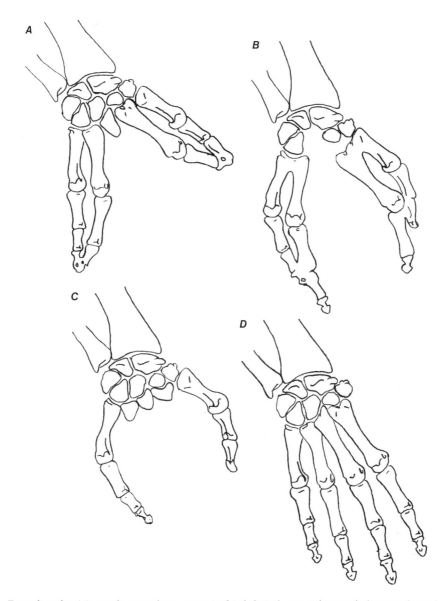

FIGURE E-6.5.2. Ectrodactyly: (*A*) nearly complete agenesis third digital ray with symphalangism lateral and medial distal phalanges; (*B*) complete agenesis third digital ray and capitate with symphalangism lateral and medial distal phalanges; (*C*) nearly complete agenesis second, third, and fourth digital rays; (*D*) single digital agenesis thumb first digital ray.

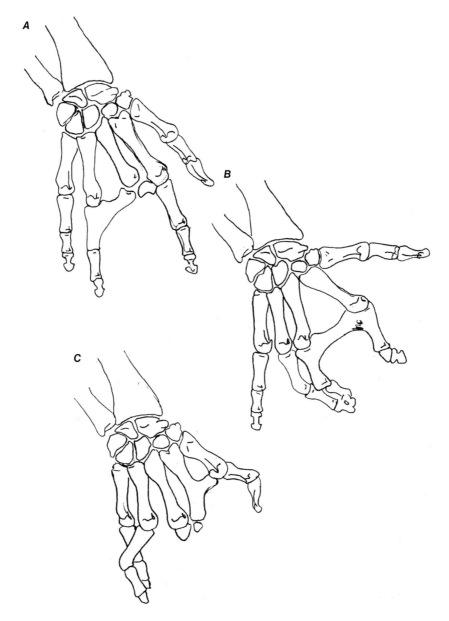

FIGURE E-6.5.3. Ectrodactyly—cross bone split hand: (*A*) affecting the third and fourth digits; (*B*) affecting the second and third digits with the distal third and fourth distal phalange syndactyly; (*C*) affecting the first, second, and third digits.

may not be associated with a similar defect in the foot. Cross bone split hand is another expression with deformed cross over proximal phalanges (Fig. E-6.5.3). There are many other variable expressions of split hand, but usually, the thumb and fifth finger digits are present. Other very rare expressions of ectrodactyly include the absence of the entire first digit alone (Fig. E-6.5.2D), or missing fourth and fifth digits with adjacent carpals similar to ulnar hemimelia but with the ulna present (Kelikian 1974; Temtamy and McKusick 1978).

E-6.6. Polydactyly

This is supernumerary or digital ray duplication. Complete duplication results in a separate, extra digital ray, while partial duplication implies bifurcation of one segment of a digital ray leading to doubling of the segments distal to it. Sometimes, a separate rudimentary digit forms, often in the soft tissue only. Incomplete duplication involves either a bifurcated shaft or bifurcated proximal articular end with duplicated distal

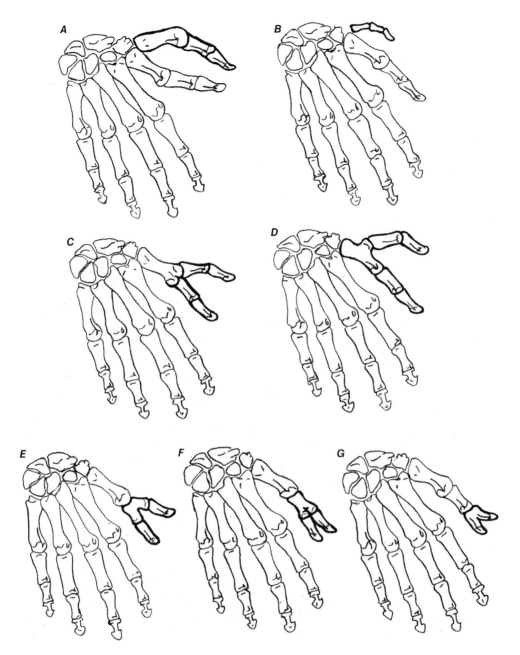

FIGURE E-6.6.1. Polydactyly—preaxial: first digit (thumb) (*A*) complete duplication; (*B*) duplicate osseous remnant; incomplete duplications; (*C*) duplicate phalanges; (*D*) branching metacarpal with duplicate phalanges; (*E*) bifurcated proximal phalange; (*F*) duplicate distal phalange; (*G*) bifurcated distal phalange.

segments. Sometimes, the entire shaft (block segment) or distal articulating end appears broader than usual (usually the metacarpal). Variations often appear in the bifurcated shaft expression, varying from Y shape to branching off at an angle.

Preaxial expressions involve the thumb and, on very rare occasions, the second digital ray (Fig. E-6.6.1).

Postaxial expressions affect the fifth digit and, on very rare occasions, the fourth digit ray (Fig. E-6.6.2).

Very rarely is the central digital ray duplicated. Preaxial polydactyly can also include duplication of the substitute finger digit in the five-fingered hand, and an additional fingerlike digit similar to the triphalangeal thumb. It is possible, but very rare, to have mixed polydactyly—both preaxial and postaxial—in the same hand.

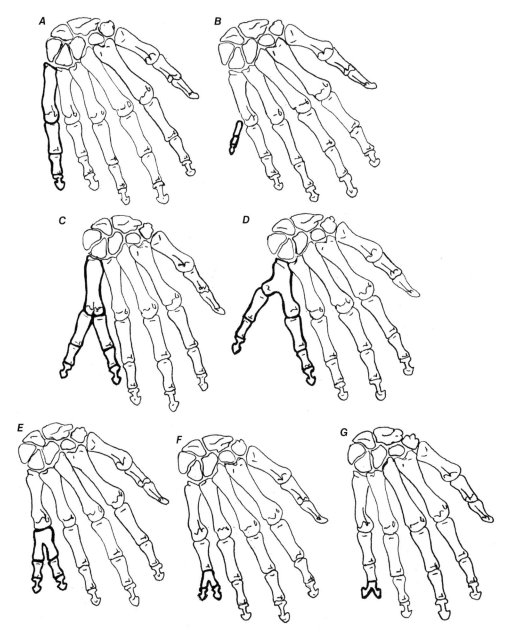

FIGURE E-6.6.2. Polydactyly—postaxial: fifth digit (*A*) complete duplication; (*B*) duplicate osseous remnant; incomplete duplications; (*C*) duplicate phalanges; (*D*) branching metacarpal with duplicate phalanges; (*E*) bifurcated proximal phalange; (*F*) bifurcated middle phalange; (*G*) bifurcated distal phalange.

CHAPTER

F

LOWER LIMBS

LOWER LIMB DEVELOPMENT

Lower limb morphogenesis lags 2–3 days behind the development of the upper limb until the last embryonic week (the eighth week). The lower limb follows the same pattern of development as the upper limb, but under the control of a different set of distinguishing genetic signals. Following the emergence of the upper limb bud from the lateral mesoderm during the third week, eruption of the lower limb bud is stimulated by an overlying ring of thickened ectoderm, the crest of Wolff, opposite the lumbar–sacral region. The underlying undifferentiated mesenchymal cells from the lateral mesoderm push outward, covered by ectodermal tissue. The bud's ventral medial border thickens to form an oblong apical ectodermal ridge (AER), the prime mover for proximal–distal growth of the limb bud. Undifferentiated mesenchymal cells directly beneath and in contact with the ridge, the proximal zone (PZ) is stimulated to multiply and advance to grow the limb bud in a proximal–distal direction. The AER and PZ work together as a functional unit via a feedback system. Cascading, overlapping genetic signals along the gradient of the lengthening mesenchymal condensation core of the limb bud define the different segments of the developing limb, beginning in the proximal region of the pelvic girdle. Genetic signals from other axes enter into a complex interaction with the proximal–distal axis with contributions to positioning and patterning of the developing limb. The anterior–posterior axis influences the development and positioning of the preaxial (tibia––first toe side) and postaxial (fibula—fifth toe side) aspects of the developing limb. The genetic signaling comes from a cluster of cells known as the zone of polarizing activity (ZPA) at the posterior border near the flank of the limb bud, adjacent to the AER. The dorsoventral axis maintains dorsoventral polarity with signals from the dorsal and ventral ectoderm. As the limb bone anlages are defined, undifferentiated mesenchyme between them (interzones) is signaled to transform into fibrous connective tissue precursors for the development of the joints.

As the hand plate takes shape during the fourth week, the caudal portion of the rounded lower limb bud tapers to a distal tip. By the fifth week, the tip transforms into the digital foot plate as the pelvic girdle, thigh, and lower leg segments are distinguished. During the sixth week, the tarsal region is outlined by condensed mesenchyme as the digital rays appear. Within a few days, the digital rays become more prominent and the margin of the foot plate becomes crenulated. The lower leg begins to flex toward the parasagittal plane from the coronal plane. The toes become distinct as cell death of interdigital tissue occurs, with plantar swellings on the tips of the toe digits during the seventh week, as the legs lengthen and the feet approach each other in midline. All primordial bones are set in hyaline cartilage by the end of the eighth week (Fig. F-1.0).

Generally, timing for development and growth is synchronized between the paired limbs so that the

Atlas of Developmental Field Anomalies of the Human Skeleton: A Paleopathology Perspective, First Edition. Ethne Barnes.
© 2012 Wiley-Blackwell. Published 2012 by John Wiley & Sons, Inc.

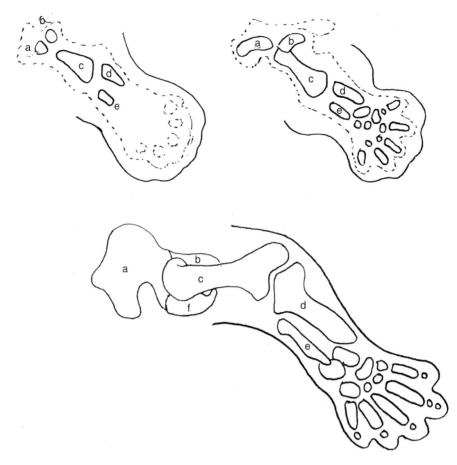

FIGURE F-1.0. Lower limb development: from 38 days (11 mm), 41 days (14 mm) to 47 days (20 mm), dotted lines show continuous mesenchymal condensations, solid lines outline the limb formation and precursor membranous bone formations—(a) ilium, (b) pubis, (c) femur, (d) tibia, (e) fibula, (f) ischium.

developed limbs are fairly equal in length. Occasionally, timing for development can be off between the whole limb or a segment of the limb, resulting in asymmetry (the right side is usually affected with shorter length more than the left side). Most often, the unevenness of the limbs is slight and unnoticed. However, lower limb asymmetry can be great enough to affect gait. Complete or incomplete duplication of limb bones is very rare except in the digits, usually with the duplicate appearing as a remnant of the other. When anomalies occur unilaterally, they most often occur on the right side. Complete absence (amelia) of the lower limb segment is very rare.

PELVIC GIRDLE SEGMENT

F-1. INNOMINATE DEVELOPMENT

This is derived from three separate bony elements—ilium, ischium, and pubis—paired with the developing sacrum from the axial skeleton. Segmentation of the

continuous mesenchymal condensation of the lengthening lower limb bud begins at the proximal end with separation into two core condensations representing the beginnings of the ilium and pubis, followed later by the ischium core after the leg segment representing the femur takes form (Fig. F-1.0). By the eighth week, the contour of the mature innominate is set in hyaline cartilage, and ossification begins in the ilium above the greater sciatic notch, followed by ossification in the ischial body and superior ramus of the pubis in the sixteenth to twentieth weeks. At birth, the three developing bony elements remain separate within the preset cartilaginous format. The primitive acetabulum remains a shallow cartilaginous cup with a triradiate stem—a Y-shaped epiphyseal plate—between the three bones forming the housing for the developing cartilaginous femoral head. The ischium and pubis rami unite around 8 years of age, followed by the union of the triradiate cartilaginous strip from one or more ossification centers, between the twelfth and sixteenth year, uniting the ilium to the other two bones. At puberty, secondary ossifications appear along the iliac crest, anterior

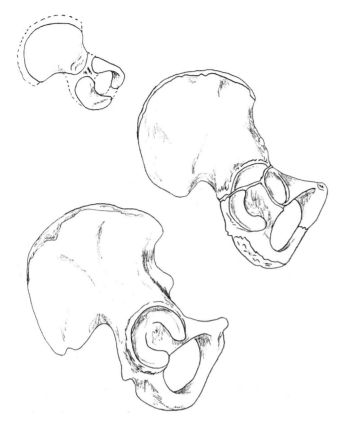

FIGURE F-1.1.0. Pelvic innominate development: from newborn bony elements—ilium, pubis, and ischium set in cartilage format to triradiate cartilage within the acetabulum of a child's innominate and mature adult's innominate.

inferior spine, ischial tuberosity, and pubic symphysis, uniting with the innominate by the twenty-third year (Fig. F-1.1.0).

INNOMINATE ANOMALIES

Dependence on the pelvic girdle for strength and stability of the entire skeleton does not allow much room for viable developmental field disturbances, unlike the shoulder girdle that can tolerate such disturbances. Developmental defects in the sacrum can affect shaping of the innominate, especially sacral agenesis associated with caudal regression syndrome.

Hypoplasia/aplasia in any of the three bony elements—ilium, pubis, and ischium—is rarely seen. The pubic symphysis can fail to develop with defective abdominal wall formation, forcing the bladder to extend outward (Ortner 2003). Bilateral agenesis of the ilium has been clinically identified in association with bilateral hypoplasia of the scapulae but is so rare that it is most unlikely to be seen in paleopathology, although the afflicted individual may survive into childhood and be

ambulatory with a waddling gait (Mac-Tniong et al. 2005).

F-1.1. Developmental Hip Dysplasia

Congenital hip dysplasia/congenital hip dislocation can be unilateral or bilateral. Developmental disturbance of the acetabulum creates instability of the hip joint leading to dislocation of the femoral head. With the coming together of the three anlage bony elements of the innominate—ilium, ischium, and pubis—a shallow basin for the primordial acetabulum takes shape in tandem with the approach of the primordial femoral head. At birth, while the acetabulum and femoral head forming the hip joint remain cartilaginous, articular cartilage receiving the femoral head within the acetabulum forms a horseshoe shape around a nonarticular fossa filled with fibrous-fatty tissue, interrupted at the inferior border by a deep notch—the acetabular notch—bridged by the transverse acetabular ligament. The depth of the acetabular socket is greatly increased by a ring of fibrocartilage—the acetabular labrum—adhering to the preosseous rim, bridging the gap formed by the acetabular notch. Pressure from the developing femoral head further deepens the socket as the labrum provides a continuous, firm suction-like hold on it, keeping the femoral head in place within the acetabulum. The small depression on the femoral head—the fovea—marking the attachment of a fibrous band—ligamentum teres—originates beneath the fibrous-fatty tissue at the inferior portion of the acetabular nonarticular fossa, from both sides of the acetabular notch. Its blood vessels provide nutrients to the developing femoral head. The capsule encasing the hip joint is attached to the entire periphery of the acetabular labrum, covers the femoral head, and most of the femoral neck, like a sleeve, reinforced by four ligaments coiling around the femoral neck, and by iliopsoas, gluteus minimus, and vastus externus muscles. Synovial membrane lines the capsule and all the structures within it, including the labrum and ligamentum teres. This tight, complex arrangement of acetabular labrum, joint capsule, associated ligaments, and muscles enforces a strong, stable joint, making it very difficult to pull apart under any circumstances (Fig. F-1.1.1A).

Developmental disturbance in this complex arrangement of the hip joint creates an unstable joint. The common denominator for the development of an unstable joint is a disturbance in the development of the acetabular labrum, particularly the superior–posterior portion, causing it to lose its suction-like hold on the femoral head. The femoral head is no longer held firmly within the acetabular socket, allowing it to be displaced (Dunn 1976; Ferguson 1968).

The defective labrum may be hypoplastic, partially aplastic, or displaced completely or partially below the

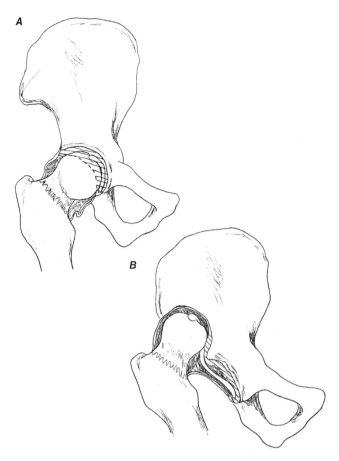

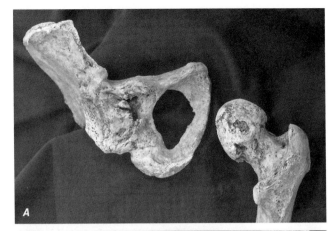

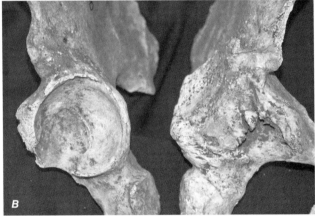

FIGURE F-1.1.1. Hip dysplasia development: (*A*) normal hip joint; (*B*) displaced hip joint (crossed lines represent labrum with the attached joint capsule outlined in double lines, ligamentum teres attached to the femoral head in solid lines).

FIGURE F-1.1.2. Developmental hip dysplasia: most likely delayed postnatal dislocation, (*A*) unilateral right hip dysplasia with proximal femur; (*B*) normal left acetabulum compared with the right atrophied acetabulum with false joint just above and behind it, adult male (NMNH 382947), SD.

acetabular rim, resulting in a variable gradient of incomplete to complete failure of the labrum's hold on the femoral head. Instability caused by a severely defective labrum appears to influence the relationship between the developing femoral head and acetabular hyaline cartilage in the neonate. Ponseti (1978) noticed that autopsied neonates with hip dysplasia exhibit a bulging ridge in the acetabular cartilage, dividing it into two parts, enabling the development of a secondary articulation superior–posterior to the acetabulum. Perhaps the movement of the femoral head out of the designated socket empowers the iliopsoas to force it up and back, placing traction on the developing acetabular cartilage (Fig. F-1.1.1B).

Most incomplete disruptions involve partial hypoplasia/aplasia of the acetabular labrum and are considered minor variants of hip dysplasia, with transient femoral head dislocation that is able to self-correct, or be corrected in early infancy with adequate interven-

tion. Some may progress upon ambulation to delayed complete dislocation (Fig. F-1.1.2).

Most neonatal complete disruptions with permanent femoral head dislocation involve the displacement of the acetabular labrum as it extends beneath the rim, stretching between the femoral head and the acetabular wall instead of extending outward along the rim, forcing the femoral head to ride up and back onto the ilium. Pressure from the femoral head against the ilium creates a false articular basin while the true acetabulum atrophies (Figs. F-1.1.3 and F-1.1.4). The ligamentum teres thickens and greatly elongates from its place of origin as it follows the displaced femoral head (Fig. F-1.1.1B). The joint capsule stretches over the relocated head, adhering to the ilium and or gluteus muscle fibers in its wake. The femur head flattens out or becomes misshapen by the new hip joint arrangement. Even with severe hip dysplasia, the dislocated hip does not cause pain but does produce a waddling gait. In time,

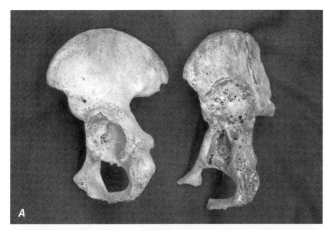

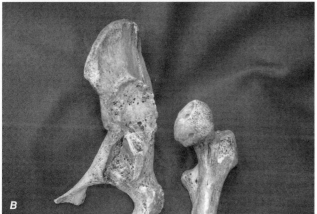

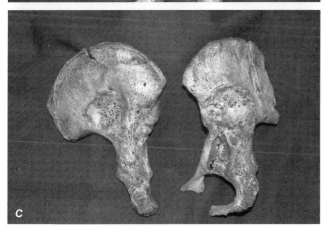

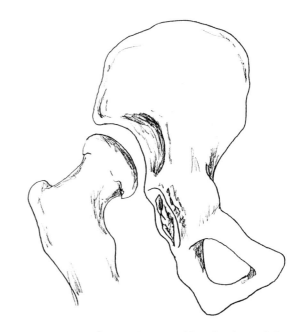

FIGURE F-1.1.4. Developmental hip dysplasia: dislocated femoral head forms false joint basin on ilium while acetabulum development fails and atrophies.

FIGURE F-1.1.3. Developmental hip dysplasia: (*A*) normal left compared with left hip dysplasia; (*B*) left hip dysplasia with proximal femur; (*C*) bilateral right and left false joint basins on ilea, adult female (NMNH 368989), Ft. Ancient, OH.

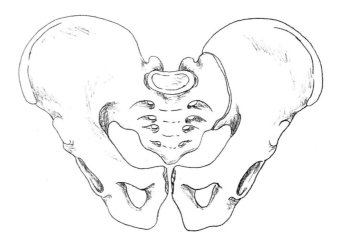

FIGURE F-1.2. Sacroiliac coalition: unilateral right.

pain comes with the onset of degenerative joint disease caused by the defect.

F-1.2. Sacroiliac Coalition

Very rarely, the facing cartilage plates of the sacroiliac joint are programmed to ossify as the bones mature, uniting the two as one. There is no indication of a joint having ever been present as the union is smooth and uninterrupted, as opposed to fusion from arthritic disease (Fig. F-1.2). This often occurs unilaterally but can also be bilateral, and causes no dysfunction.

THIGH SEGMENT

F-2. FEMUR DEVELOPMENT

Ossification begins in the shaft by 28 days within its hyaline cartilage anlage. The shaft is completely ossified at birth, while the proximal neck and head remain cartilaginous. The distal end is also cartilaginous at birth but contains an ossification center that appears at the end of the last trimester. Ossification progresses from the shaft into the neck while the head ossifies from its own center, beginning around the sixth postnatal month. Separate ossification centers for the greater trochanter appear around the fourth year, and for the lesser trochanter by the eleventh year. The ossified head unites with the femoral neck during adolescence usually by 18 years, followed by the trochanters. The distal epiphysis is the last to unite with the shaft by around the twentieth year.

FEMUR ANOMALIES

F-2.1. Proximal Femur Variations

These usually involve variable angles of the femoral neck.

Asymmetrical Torsion of the Femoral Neck
This is the most commonly occurring variation of the proximal femur, with one side at a normal angle to the shaft with the other side showing a decreased angle to the shaft (Fig. F-2.1A). This does not appear to cause disturbance in gait.

Hypoplasia of the Femoral Head and/or Neck
This can occur especially with hip dysplasia. Shortening of the neck with or without a small femoral head dramatically reduces the angle of inclination with the shaft (Fig. F-2.1C), thus disrupting ambulation.

Coxa Vara
This is a decreased inclination of the angle of the femoral neck that has the greater trochanter prominently appearing above the level of the femoral head, as the neck droops downward, often associated with a shortened or bowed femur (Fig. F-2.1D). Coxa vara causes stiffness and a marked painful limp with limited abduction that can be disabling if bilateral.

Coxa Valga
This is an increased inclination angle of the femoral neck that places the femoral head high above the greater trochanter as the neck grows nearly straight

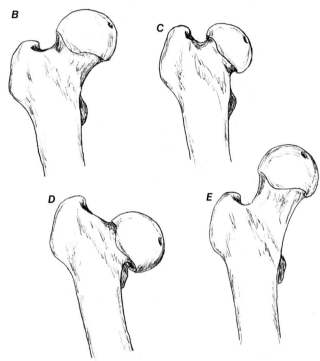

FIGURE F-2.1. Proximal femur variations: (*A*) asymmetrical torsion of femoral necks; (*B*) normal proximal femur; (*C*) hypoplasia of the femoral neck and head; (*D*) coxa vara; (*E*) coxa valga.

upward (Fig. F-2.1E). It holds the leg in stiff abduction with limited adduction, interfering with ambulation.

F-2.2. Femur Hypoplasia/Aplasia

This usually affects the proximal end of the femur, leaving the distal end developing as programmed, referred to as proximal femoral focal deficiency with a wide range of variable expressions (Gragon et al. 1987).

Proximal Femoral Focal Deficiency
The primary developmental disturbance appears to be within the primordial mesenchymal femoral neck as the proximal portion of the femur emerges in the mesenchymal lower limb core below the innominate segment, before extending into the lower portion of the femur segment and beyond. This in turn disturbs the developing hyaline cartilage and ossification processes of the

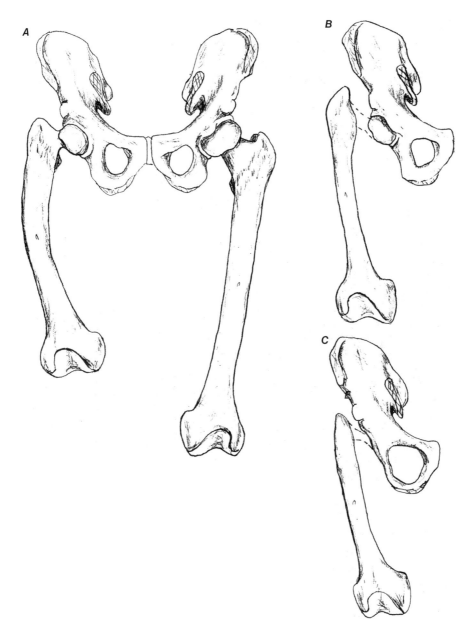

F I G U R E F -2.2. Femur hypoplasia/partial aplasia (proximal femoral focal deficiency): (*A*) simple hypoplasia with coxa vara; (*B*) bony aplasia of the neck with the fibrocartilage band (dotted lines) connected to the bony head in the acetabulum, proximal end above the acetabulum; (*C*) bony aplasia of the neck and head with the fibrocartilage band (dotted lines) connecting the proximal end to the aplastic acetabulum.

head and neck, resulting in either simple hypoplasia of the femur, or disrupted ossification of the neck, and sometimes the head as well with a much shortened femur.

Coxa vara often accompanies simple hypoplasia of the femur with a shortened shaft that bows inward. Femur shortening usually does not exceed 3 in. of expected length, with the leg fixed in external rotation, and the patella is usually smaller (Fig. F-2.2A).

More severe expressions of the proximal femoral focal deficiency leave the neck as a fibrocartilage band

connected to the femoral head instead of ossifying. Occasionally, this band will ossify. The femoral head may completely ossify as expected and be contained within the acetabulum, ossify as a remnant femoral head within a remnant acetabulum, or not develop at all with aplasia of the acetabulum. The proximal end of the femur is tapered and locates above the level of the acetabulum. With aplasia of the entire hip joint, the obturator foramen enlarges. The femoral shaft narrows and the leg can be up to 6 in. shorter than expected (Fig. F-2.2B,C). Needless to say, ambulation is greatly affected with the femur in fixed flexion deformity and limited or absent hip abduction. Disturbance in the development of the patella, lower leg, and or foot is often associated with this type of defect (Gillespie and Torode 1983).

Phocomelia

Aplasia of the femur or lower leg, or both, with vestiges or complete foot plate development attaching to the remaining leg part or the pelvis, unlike phocomelia in the upper limb, is highly unlikely to be seen in paleopathology.

F-2.3. Bifurcated Distal Femur

This is very rare and usually occurs with paraxial hemimelia—fibular or tibial agenesis—suggesting embryonic paraxial disturbances influencing distal femur development. While the proximal femur end develops as expected, the shaft flattens out and the distal portion divides into an upside-down Y shape (Fig. F-2.3) with the patella usually absent. Most often, it is the preaxial (tibial) side that is absent, leaving an enlarged fibula articulating with one end of the bifurcation, with ankle

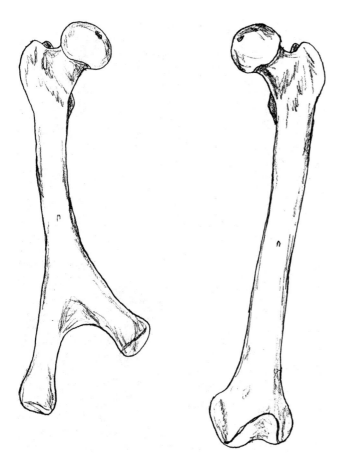

FIGURE F-2.3. Bifurcated distal femur.

and foot defects. The lower leg segment is usually in fixed flexion (Aalami-Harandi and Zahir 1976; Wolfgang 1984). Ostrum et al. (1987) reported a unilateral bifurcated femur with a normal lower leg and foot segment. Apparently, the bifurcation was the only defect present in this individual with little disturbance to ambulation. The tibia articulated with one end of the bifurcation.

F-3. PATELLA DEVELOPMENT

This is the largest sesamoid bone in the body and develops within the quadriceps femoris tendon as it passes over a depression on the ventral distal femur between the condyles to attach via the patellar ligament to the tibial tuberosity. While the tendonous fibers attach to the ventral surface, the dorsal surface and adjacent femoral surface are covered with articular cartilage and are contained within the knee joint capsule (Fig. F-3.1A). Embryonic patella development is regulated by the dorsoventral axis. The patella protects the femoral ventral articular cartilage of the knee joint and increases the leverage for the extension of the quadriceps. The cartilaginous precursor of the patella appears by the third fetal month. Ossification usually begins between 2 and 5 years with the appearance of several ossifying nuclei programmed to quickly coalesce into one central ossification center. Completed ossification occurs immediately after puberty.

PATELLA ANOMALIES

F-3.1. Patella Hypoplasia/Aplasia

This is often bilateral. The sulcus between the femoral condyles may deepen, and the patellar articular surface is absent with patellar agenesis (Fig. F-3.1C). Extension of the knee is hampered with the absence of the patella, making it difficult to stand upright and walk (Jerome et al. 2009). A very small patella is less likely to interfere with knee function (Fig. F-3.1B). Very rarely, a hypoplastic patella can be developmentally displaced laterally, resulting from either aplasia of the lateral femoral condyle or abnormal alignment of the quadriceps femoris tendon. Genu algum (knock knee) often results with the inability to laterally rotate the knee, as well as inability to extend the knee with the potential for a fixed flexion deformity (Green and Waugh 1968).

F-3.2. Segmented Patella

Occasionally, not all of the osseous nuclei coalesce into one ossification center; instead they form one or more separate centers that result in the segmentation of the

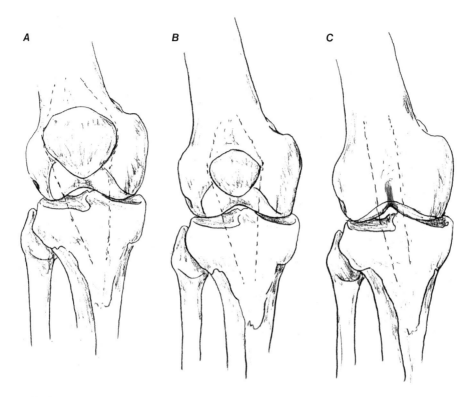

FIGURE F-3.1. Patella hypoplasia/aplasia: (quadriceps femoris tendon/ligamentum patella represented by broken lines) (*A*) normal patella; (*B*) patella hypoplasia; (*C*) patella aplasia.

patella (Fig. F-3.2.1). Usually, only one separated ossification appears, known as bipartite patella (Figs. F-3.2.2 and F-3.2.3). More than one separated ossification is a rarity. The separated ossifying segment may eventually completely or partially unite with the mature body patella, remain a separate ossicle segment connected to the body patella by fibrocartilage, or simply not ossify at all. The superior–lateral margin is primarily targeted, very rarely involving the medial or inferior margins. The superior–lateral vastus notch (sometimes known as the emarginated patella) represents lack of ossification of a segment. Vastus notch presents with a smooth border as opposed to the rough bony interface of the fibrocartilaginous connection between the bipartite separated ossicle segment and the body patella.

Duplicate patella (Fig. F-3.2.1I) evolves from the duplication of the precursor patella segment and is extremely rare, with the secondary, smaller bone usually riding behind the primary patella (Brailsford 1948).

LOWER LEG AND FOOT SEGMENTS

PARAXIAL DEVELOPMENT

The lower leg segment containing the tibia and fibula also includes the proximal talus and calcaneus tarsals

of the hindfoot that are closely aligned with the developing distal tarsals of the midfoot from the foot plate along with the forefoot digital rays. Similar to upper limb development, the anterior–posterior axis influences the development along the proximal–distal axis of the lower leg and foot segments with polarity determined by the dorsoventral axis. This complex interaction of genetic signaling is responsible for the development and positioning of the separate bones in the lower leg segment—tibia, fibula, proximal tarsals, foot segment distal tarsals, metatarsals, and digital phalanges. Development of the tibia, talus, navicular, first and second cuneiforms, first and second metatarsals, and digital phalanges occur along the anterior (preaxial) axis, while the posterior (postaxial) axis influences the development of the fibula, calcaneus, cuboid, fifth and fourth metatarsals, and digital phalanges (Fig. F-4.0). The central ray—third cuneiform and third metatarsal and digital phalanges—separating the two sides can be influenced by either one but most often follows the preaxial signals.

While the hand spreads out and requires considerable dexterity, emphasis on stability for body weight bearing on the ankle and foot creates a far different configuration of its parts. As the proximal preaxial tarsal—the talus—separates from the preaxial side of the lower leg, it moves from a precartilaginous parallel position with

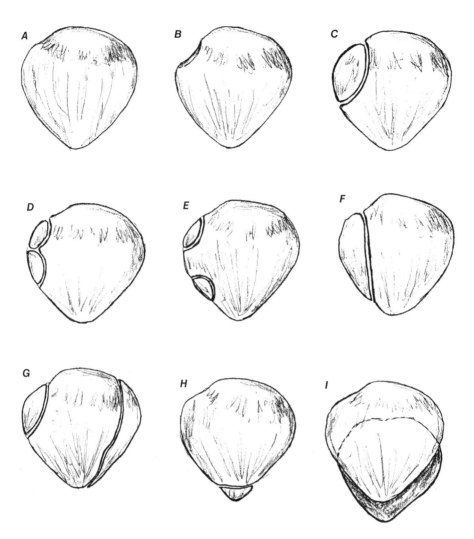

FIGURE F-3.2.1. Segmented patella: (right ventral) (*A*) normal; (*B*) vastus notch; (*C*) bipartite; (*D–H*) very rare atypical variations; (*I*) very rare duplicate patella.

FIGURE F-3.2.2. Segmented bipartite patella: (dorsal) unilateral left bipartite and normal right, adult male (NMNH 308677), Hawikuh, NM.

FIGURE F-3.2.3. Segmented bipartite patella: (ventral) bilateral, adult male, Byzantine Petras, Crete.

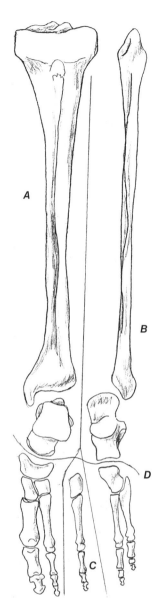

FIGURE F-4.0. Lower leg and foot paraxial development: (a) preaxial (tibial) side—tibia, talus, navicular, first and second cuneiforms, and digital rays; (b) postaxial (fibular) side—fibula, calcaneus, cuboid, fifth and fourth digital rays; (c) central axis third cuneiform and digital ray; (d) transverse line divides the lower leg segment and foot segment.

the budding proximal postaxial tarsal separating from that side of the lower leg—the calcaneus—to sit on top of it, causing the inner side of the foot to arch longitudinally and transversely. The arches allow the foot to absorb and redistribute superimposed body weight, while the outer side of the foot maintains positioning when the foot is pressed to the ground. This configuration of bones along with interosseous ligaments resembles a twisted plate that can untwist and twist with changes in load bearing for ambulation.

F-4. TIBIA AND FIBULA DEVELOPMENT

These are closely aligned as one unit for strength and stability, connected by interosseous membrane. The much larger tibia carries the weight of the body, takes part in the knee joint, and rests on top of the talus at the ankle joint, while the slender fibula acts as a strut, absorbing the shocks of weight bearing on the tibia. Both have distal bony extensions—lateral fibular malleolus and medial tibial malleolus—bracing the ankle within a mortise joint. Ossification for the shaft of the tibia begins in the seventh week, followed by ossification beginning in the shaft of the fibula. The ossification center for the proximal epiphyses of the tibia appears during the last fetal month with ossification extending downward to include the tuberosity between 8 and 14 years, and the epiphyses unites with the shaft by the twentieth year. The ossification center for the distal tibial epiphyses appears at the end of the first year, uniting with the shaft around 18 years. The distal fibular epiphysis appears by the second year, uniting with the shaft by 19 years, while the proximal fibular epiphysis appears between 3 and 5 years, uniting with the shaft by 20 years.

TIBIA AND FIBULA ANOMALIES

F-4.1. Lower Leg Meromelia (Congenital Amputation)

Transverse limb deficiency of the lower leg is extremely rare. This represents failure of the mesenchymal limb core to reach its full extent, cutting off the development below a certain level, usually below the knee, or at the end of the lower leg segment with developmental failure of the foot plate (Fig. F-4.1). Failure in development can also reach to the foot plate.

F-4.2. Lower Leg Paraxial Hemimelia

This is the agenesis of the tibia or fibula. This can be a complete or partial absence of the tibial (preaxial) side or the fibular (postaxial) side, and is most often unilateral. The missing bone is represented by a fibrocartilaginous band, suggesting arrested development of embryonic tissue for the absent bone. Development of associated paraxial tarsals and digital rays of the affected side is usually disturbed as well, and the femur is often shortened somewhat, sometimes with hypoplasia of the distal condyles, and a small or absent patella. There is no simple formula for lower leg paraxial hemimelia disorders, with a wide range of variable compositions for either side of the lower limb segment (Fig. F-4.2.0),

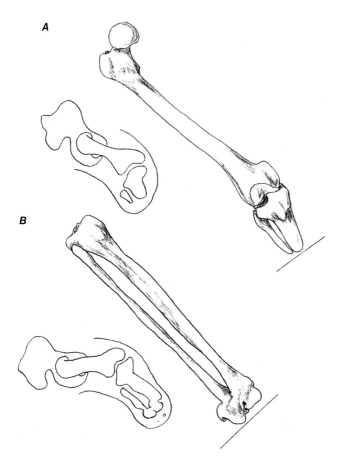

FIGURE F-4.1. Lower leg meromelia (congenital amputation): transverse limb deficiency: (*A*) below the knee; (*B*) distal end of the lower leg segment.

but there have been many attempts to classify the different expressions. Walking with a painless limp is usual, and the uneven leg lengths promote vertebral scoliosis. Bilateral expressions do allow the individual to ambulate on shortened legs, albeit awkwardly (Gragon et al. 1987).

Tibial (Preaxial) Hemimelia

This is much less common than fibular (postaxial) hemimelia. Generally, the fibula thickens as it takes the body weight meant for the tibia (Fig. F-4.2.1), and dislocates upward and back to articulate with the posterior–upper aspect of the lateral femoral condyle. The fibula may be bowed to some degree. Often, the talus and calcaneus of the hindfoot are united, sometimes uniting with the navicular. The navicular and cuneiforms are often affected along with the first, second, and sometimes third digital rays, with all or some of them either consolidating or absent, or they may be normal. Most often, the first digital ray is absent, sometimes including the navicular. The foot position presents as talipes equinovarus club foot, with the plantar surface directed

inward, and in severe cases, the foot is also supinated so that it points upward (talipes calcaneus). Preaxial polydactyly is common. Partial agenesis of the tibia may leave either the proximal or distal end in place, either of nearly normal size or as a remnant. Most often, it is the proximal end remaining in place (Fig. F-4.2.1B). Sometimes the distal portion of the tibia tapers to an end that diverges away from the distal fibula. Ambulation with bilateral absence of the tibia, albeit with short legs, has the fibulae abducted and rotated outward and marked talipes equinovarus club feet.

Fibular (Postaxial) Hemimelia

This is more common than tibial (preaxial) hemimelia. The tibia is usually shortened with the midportion bowed anteriorly and thickened cortex on the concave side. The amount of bowing varies, or the tibia can be of normal size and shape. The inferior malleolus may be knobby with an oblique articulation for the ankle joint. The ankle is usually deformed in talipes equinovalgus club foot (Fig. F-4.2.2). The heel of the calcaneus is elevated and turned outward, and may be displaced posterior laterally to the tibia with the talus medially and distally to it. A ball-and-socket ankle joint may prevail instead of the usual hinge joint. Union of the hindfoot calcaneus and talus is common, sometimes includes union with other tarsals, or the tarsals may be deformed. The fifth and fourth digital rays may be absent; sometimes the cuboid is also absent. The fifth digital ray may be the only absence in the foot, or the foot may be normal with an adequate ankle joint. Partial absence of the fibula usually leaves the lower part present high above the ankle joint, thus not allowing it to participate in the ankle movements.

F-4.3. Duplication (Dimelia) Lower Leg Ray

This is extremely rare and usually involves duplication of the postaxial fibula with agenesis of the preaxial tibia and associated foot rays. The two fibulae face each other, with the medial one usually shorter than the lateral one. The postaxial fibular rays in the foot duplicate as well, with the duplicates on either side of a central digit, resulting in a "mirror foot" with seven or more digits (Fig. F-4.3).

F-4.4. Tibia–Fibula Synostosis

This is similar to radial–ulnar synostosis but much less common and far less complex. The superior tibial–fibular joint has only minimal gliding movement, designed to absorb body weight shocks, unlike the superior radial–ulnar joint involved in rotational movements of the forearm. During embryogenesis, the

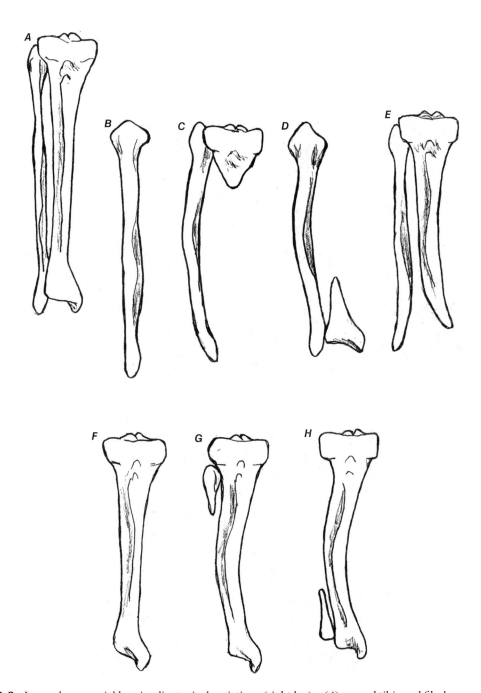

FIGURE F-4.2.0. Lower leg paraxial hemimelia: typical variations (right leg)—(*A*) normal tibia and fibula; preaxial; (*B*) complete tibia agenesis with a wide fibula; (*C*) incomplete tibia agenesis with the proximal end present; (*D*) incomplete tibia agenesis with distal end present; (*E*) distal tibial end agenesis with divergence distal ends; postaxial; (*F*) shortened tibia with complete fibula agenesis; (*G*) incomplete fibula agenesis with the proximal end present; (*H*) incomplete fibula agenesis with the distal end present.

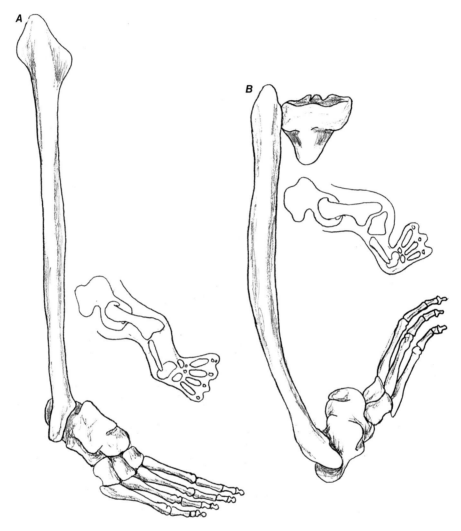

FIGURE F-4.2.1. Tibia (preaxial) hemimelia: (right) (*A*) complete tibia agenesis with calcaneus–talus–navicular coalition, agenesis first cuneiform and first digital ray, talipes equinovarus foot; (*B*) incomplete agenesis with proximal tibia present, calcaneus–talus coalition, agenesis navicular, first and second cuneiforms and digital rays, talipes calcaneus foot.

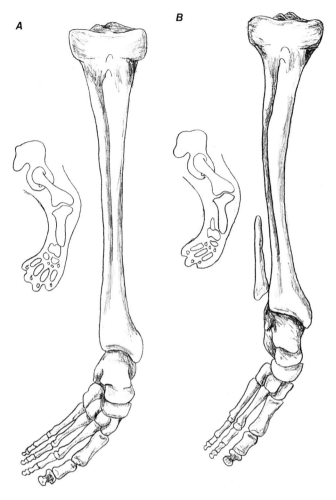

FIGURE F-4.2.2. Fibula (postaxial) hemimelia: (right) (*A*) complete fibula agensis with calcaneus-talus coalition, agenesis fifth digital ray, talipes equinovalgus foot; (*B*) incomplete agenesis with distal fibula present, calcaneus-talus coalition, absent cuboid, fifth and fourth digital rays, talipes equinovarus foot.

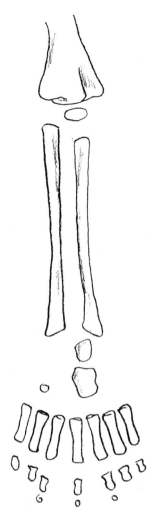

FIGURE F-4.3. Duplication (dimelia) fibular postaxial ray: agenesis preaxial tibia and first digital ray, mirror duplication remaining digital rays on both sides of the central digital ray (infant).

interzone required for the joint to develop may fail to appear as the separation of the tibia and fibula anlages progresses from the distal end to the proximal end of the mesenchymal core, allowing preosseous cartilage tissue to fill the space and ossify (Fig. F-4.4). Occasionally, the distal joint may fail to develop.

F-5. TARSUS DEVELOPMENT

Seven closely packed irregularly shaped bones are separated by synovial joints and held together by interosseous ligaments. The proximal tarsals of the hindfoot—talus and calcaneus—developing from the distal end of the lower leg segment, join the remaining tarsals of the midfoot from the proximal margin of the foot plate—navicular, the three cuneiforms, and cuboid—that joins

with the forefoot formed by the digital rays—the metacarpals and phalanges. The articulation between the talus and distal tibia along with the fibular malleolus forms a mortise synovial hinge joint (talocrural joint) for the ankle. The hindfoot and midfoot sections are separated by the compound transverse tarsal joint consisting of the talonavicular and calcaneocuboid joints (Fig. F-5.0). The navicular is involved in both hindfoot motion and compensatory midfoot motion, acting as a bridge between the two.

The individual tarsals vary in size and shape within a set pattern of development as they facilitate multiple articulations with each other. Each evolves from mesenchymal nuclei within the embryonic tarsal region, consolidating into a single preosseous cartilaginous center preceding ossification. Rarely, more than one preosseous center develops, often consolidating into

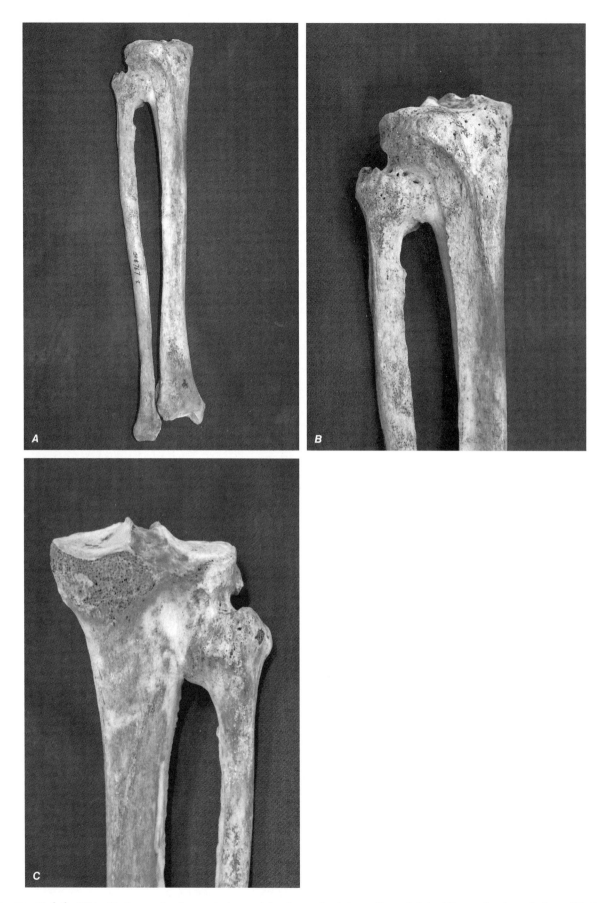

FIGURE F-4.4. Tibia–fibula proximal synostosis: conjoined superior joint unilateral right (*A*) anterior–lateral view; (*B*) close-up; (*C*) close-up dorsal view, adult female (NMNH 308767), Hawikuh, NM.

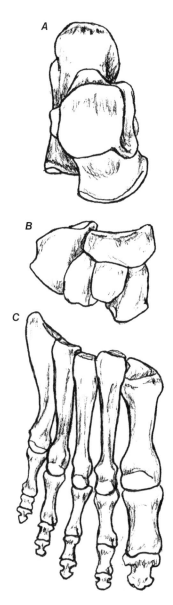

FIGURE F-5.0. Foot divisions: (*A*) hindfoot—calcaneus and talus tarsals; (*B*) midfoot—navicular, cuneiforms, and cuboid tarsals; (*C*) forefoot—metatarsals and phalanges.

one final osseous center. Separated preosseous nuclei can develop into small extra ossicles within the tarsus, but are usually unidentifiable in paleopathology. At birth, the only bony representations of the tarsals are the calcaneus, talus, and often the beginning cuboid, while the others remain completely cartilaginous. The ossification center for the calcaneus appears between the fourth and sixth fetal month, followed by the talus around the seventh fetal month, and the cuboid ossification center appears around the time of birth. Ossification of the third cuneiform begins during the first year, followed by the first cuneiform in the third year, the second cuneiform during the fourth year along with the

navicular. Secondary ossifying epiphysis for the dorsal aspect of the calcaneus develops between 7 and 10 years, uniting with the bone following the onset of puberty.

TARSAL ANOMALIES

F-5.1. Club Foot (Talipes Equinovarus)

This is a twisted foot that accounts for about 95% of identified club feet. Variable expressions range from minor to major, with about half appearing bilateral (Castriota-Scanderbeg and Dallapiccola 2005). The primary default lies within the hindfoot setting off a chain of compensatory reactions in the rest of the foot. The hindfoot turns inward (varus), raising and shortening the inner border (supination), sometimes forcing it abnormally high (cavus), with the midfoot and forefoot turned inward (adducted) and often extended downward (plantar flexed). The affected tarsals are usually smaller than normal. While minor variable combinations of midfoot and forefoot reactive anomalies occur, the major hindfoot anomalies are remarkably consistent (Settle 1963). Sometimes, the ankle mortise is externally rotated, or a ball-and-socket joint develops.

The key defect lies within the developing embryonic precursor talus as it separates from the preaxial lower leg segment before it locates over the precursor calcaneus. Development appears disorganized, resulting in an abnormal bone that is nearly unrecognizable as a talus (Fig. F-5.1A–C). The most severe distorted development takes place in the talar neck—shortened and grossly distorted, or absent—pulling a flattened talar head medially and downward.

A single, distorted, and medially slanted subtalar articulation forms, forcing the calcaneus to rotate inward into an equinovarus position. The sustentaculum tali of the calcaneus often rest against the tip of the medial malleolus of the tibia. Articulations with the midfoot cuboid and navicular are displaced medially and plantarward on top of each other instead of side by side, twisting the foot so that weight bearing is placed along the upper, outer side of the foot (Fig. F-5.1E).

The lateral margin of the displaced cuboid becomes rounded, and as the plantar surfaces of both bones are wedged together, arching of the shortened inner border increases in height. The navicular articulates with the medial aspect of the disrupted talar head, and sometimes articulates with the anterior medial malleolus of the tibia, and/or with the posterior aspect of the lateral malleolus of the fibula. The navicular tuberosity is often enlarged, and the displacement of this bone may place it in contact with the sustentaculum tali of the calcaneus and medial margin of the talus. The wedge-shaped

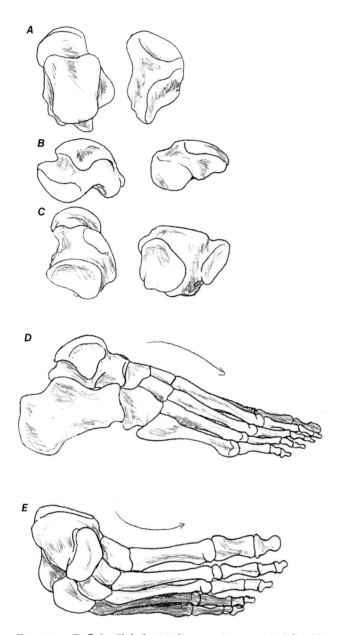

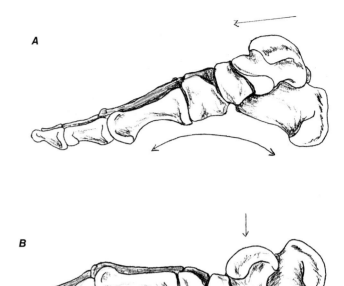

FIGURE F-5.2. Vertical talus: (*A*) normal orientation of the talus; (*B*) "rocker bottom" foot with straight arrows following vertical displacement of the talus.

F-5.2. Vertical Talus

This is a rare anomaly but known to be familial, with the talar head displaced downward and medially on the calcaneus during development with lengthening of the talar body, forcing the navicular to articulate with the neck of the talus (Fig. F-5.2B). The dislocated talar head flattens, and a dorsal facet appears on the talar neck for articulation with the navicular. With the talus locked into an abnormal vertical position, the calcaneus is forced into an equinus plantar-flexed position, with the anterior surface deviated laterally. The anterior subtalar facet is usually absent with hypoplasia of the middle facet, and the posterior facet abnormal, the articulation fixed. The cuboid may also be dislocated on the calcaneus, and other bones may appear out of alignment. The plantar surface of the foot takes on a rigid "rocker bottom" shape with the forefoot abducted and dorsiflexed. Calluses form underneath the talar head as it appears on the medial sole of the foot. The rigidity of the hindfoot and abnormal positioning of the forefoot make it progressively difficult and painful to ambulate, eventually impossible (Greenberg 1981; Viladot 1988).

F-5.3. Tarsal Coalitions

This is a union between two or more tarsals and mostly occurs from the developmental failure of the embryonic

FIGURE F-5.1. Club foot (talipes equinovarus): (right) (*A*) superior view of a normal and abnormal talus; (*B*) medial view of a normal and abnormal talus; (*C*) inferior view of a normal and abnormal talus; (*D*) lateral view of a normal foot; (*E*) lateral view of a talipes equines foot.

cuneiforms contract inferiorly with the tighter packing caused by forced adduction from the rotated tarsals, and the metatarsals, also tightly packed, rotate inwardly along with the corresponding tarsals (Settle 1963; Wright 2011). Those bones forced into abnormal weight bearing along the side and top of the lateral bones thicken and undergo bone changes from bony callus buildup over time from walking on the side or top of the foot, often with the cuboid acting as pseudo heel (Brothwell 1967).

interzone between adjacent tarsals. Union can also develop with formation of a mesenchymal connecting bridge during morphogenesis. About half of coalitions occur bilaterally. The union can be completely or incompletely osseous, or the precursor interzone or bridge remains fibrocartilaginous (nonosseous) as reflected by pitting lesions on affected interfacing tarsal surfaces. The majority of tarsal coalitions are between the calcaneus and navicular (Fig. F-5.3.1), talus and calcaneus (Fig. F-5.3.2), followed by a much lesser frequency in order: talus–navicular, calcaneus–cuboid, cuboid–navicular, navicular–first cuneiform (Figs. F-5.3.3 and F-5.3.4), and least of all between cuneiforms. Rarely, more than one set of tarsals are united in the same foot, and very rarely, mass coalition of the tarsus can occur.

The calcaneus and navicular usually do not articulate with each other. During morphogenesis, hypoplasia

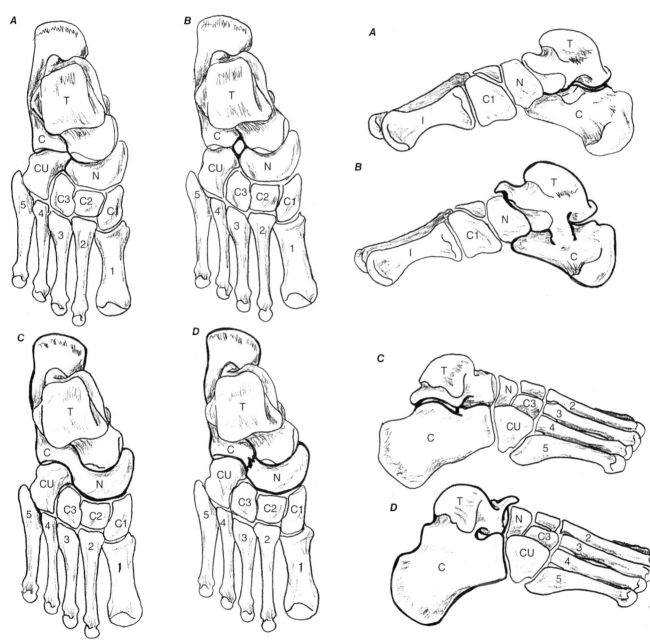

FIGURE F-5.3.1. Calcaneus–navicular coalitions: (right) (*A*) normal; (*B*) ossicle between the calcaneus and navicular; (*C*) united calcaneus–navicular; (*D*) incomplete union of the calcaneus–navicular. C = calcaneus; T = talus; N = navicular; CU = cuboid; C1–C3 = cuneiforms.

FIGURE F-5.3.2. Talus–calcaneus coalitions: (right) (*A*) normal medial view; (*B*) bony bridge uniting subtalar middle articulation with talar beaking over the navicular; (*C*) normal lateral view; (*D*) union at the subtalar posterior articulation, talar beaking over the navicular. C = calcaneus; T = talus; N = navicular; CU = cuboid; C1 and C3 = cuneiforms.

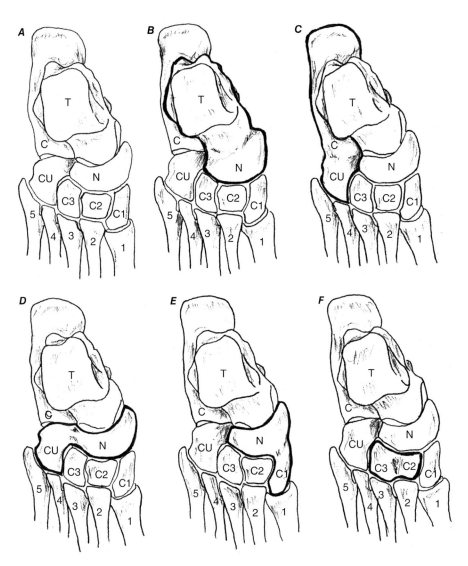

FIGURE F-5.3.3. Tarsal coalitions: (right) (*A*) normal; (*B*) talonus–navicular; (*C*) calcaneus–cuboid; (*D*) cuboid–navicular; (*E*) navicular–first cuneiform; (*F*) second cuneiform–third cuneiform. C = calcaneus; T = talus; N = navicular; CU = cuboid; C1–C3 = cuneiforms.

of the talar head, even slightly, can leave a space between the developing anterior medial border of the calcaneus and the medial distal border of the navicular that may be bridged by mesenchymal tissue to maintain uniformity. This in turn develops into a fibrocartilaginous bridge between the two that is apparent at birth, and often ossifies along with the navicular (Griffin and Rand 1988). Sometimes, a separate ossicle forms within the bridge space, or the connection remains nonosseous, reflected in pitted articulations between them (Silva 2005) (Fig. F-5.3.1).

Most coalitions of the talus and calcaneus occur at the middle subtalar articulation for the talus and the calcaneal sustentaculum tali, in the form of a bridge between the two that forms in the mesenchymal tissue during morphogenesis. This connection, bony or fibrocartilaginous, can be outside the articular surfaces or can include the articular surfaces (Kawashima and Uhtoff 1990). With this disturbance, the talar lateral process develops broader and rounder than usual. This union limits subtalar movement, forcing the navicular to pass over the talar head in compensation, and can cause a rigid, painful foot. Dorsiflexion of the foot elevates and stretches the talocalcaneal ligament that can provoke an osseous response with the formation of bony beaking projecting from the dorsal surface of the talar head. Union at the subtalar anterior or posterior articulations is rare (Viladot 1988) (Fig. F-5.3.2).

F-5.4. Tarsal–Metatarsal Coalitions

Osseous or nonosseous union can occur between any metatarsal and adjacent tarsal, but most often appears between the third cuneiform and third metatarsal of the

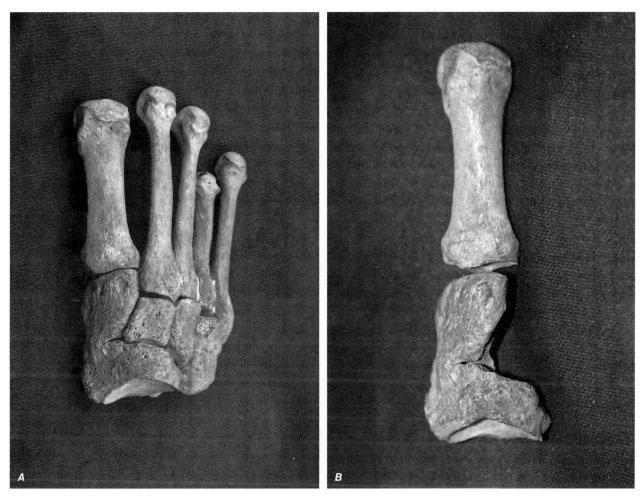

FIGURE F-5.3.4. Navicular–first cuneiform coalition: (*A*) right foot (note short MT4-brachydactyly); (*B*) close-up united tarsals with MT1, adult (NMNH 327128), Pueblo Bonito, NM.

middle digital ray. Usually, this union is incomplete nonosseous (Figs. F-5.4.1 and F-5.4.2), presenting a plantar fibrocartilaginous connection between the inferior one-third of the bony interfaces, reflected in pitting lesions (Regan et al. 1999).

F-5.5. Metatarsal–Phalanx Coalitions

This is a very rare union between a proximal phalanx and associated metatarsal of the same digital ray (Fig. F-5.5), mentioned here with the other foot coalitions instead of with digit anomalies. This is most often osseous but may be fibrocartilaginous joining similar to other nonosseous coalitions.

F-5.6. Tibia–Hindfoot Coalition

This is a rare union of the distal tibia with the talus as the tibiotalar joint fails to develop. Often, the coalition includes the calcaneus (Fig. F-5.6); this very rarely involves all of the tarsals.

F-5.7. Tarsals Bipartite and Separate Marginal Elements

These are derived from separated ossifying centers during morphogenesis. Separated marginal elements may in time unite completely or partially with the parent bone, or remain as a separate ossicle that usually attaches to the primary bone by fibrocartilaginous tissue. Often, separated marginal elements are lost but can be identified by the void left behind and by the reaction to their attachment on the surface of the parent bone. Separated marginal elements have been identified with the navicular tuberosity as the accessory os navicular (Figs. F-5.7.1A and F-5.7.2), the lateral (posterior) tubercle of the talus as the os trigonum (Figs. F-5.7.1B and F-5.7.3), and the anterior articular area of the sustentaculum tali of the calcaneus as the os calcaneus secundarius (Figs. F-5.7.1C and F-5.7.4). Known tarsal divisions (complete and partial) seem to be limited to the first cuneiform and navicular (Fig. F-5.7.1D,E), generally united to each other by fibrocartilage. Partition of the first cuneiform is a horizontal division into dorsal

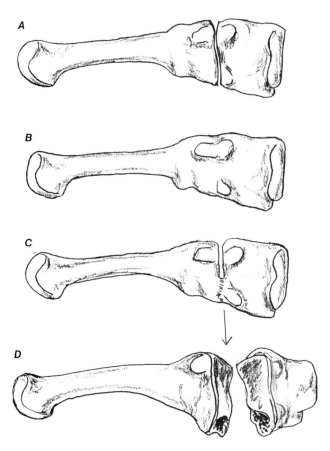

FIGURE   F-5.4.1. Tarsal–metatarsal   coalition   MT3–cuneiform: (right): (*A*) normal; (*B*) complete osseous union; (*C,D*) nonosseous fibrocartilaginous union.

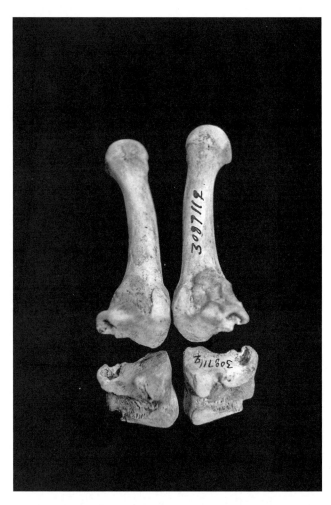

FIGURE F-5.4.2. Nonosseous coalition MT3–cuneiform: bilateral, adult female (NMNH 308711), Hawikuh, NM.

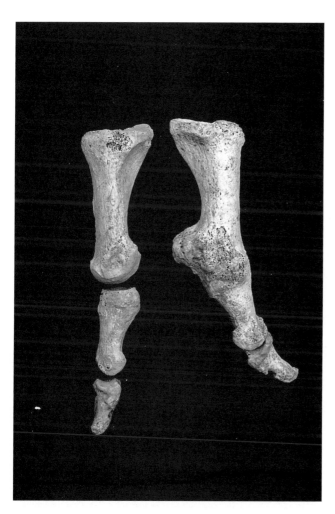

FIGURE F-5.5. Metatarsal–phalanx coalition: normal right first digit and left united MT1-proximal phalanx, adult female, Frankish Corinth, Greece.

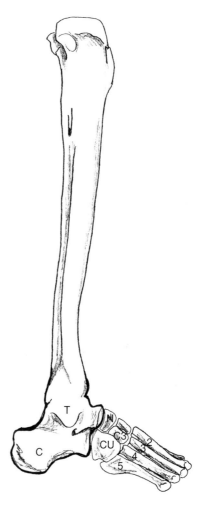

FIGURE F-5.6. Tibia–hindfoot coalition: (right) tibia, talus, and calcaneus united, right side. C = calcaneus; T = talus; N = navicular; CU = cuboid; C3 = cuneiform.

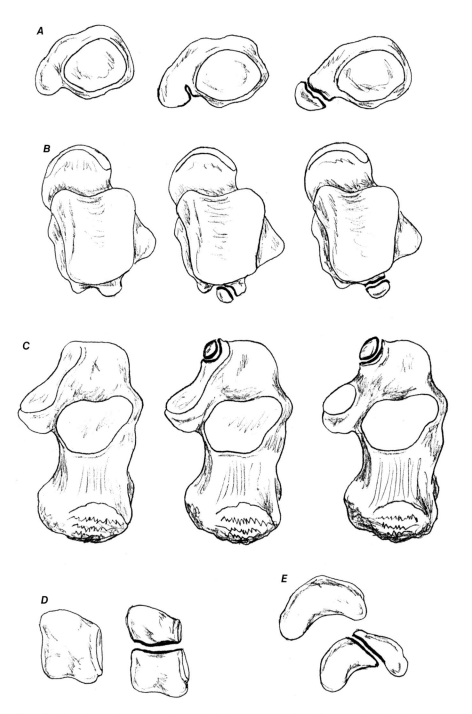

FIGURE F-5.7.1. Tarsals bipartite and separated marginal elements: (right) (*A*) normal navicular with partial bifurcation and complete os navicular; (*B*) normal talus with partial and complete os trigonum; (*C*) normal calcaneus with two expressions of calcaneus secundarius; (*D*) normal and bifurcated first cuneiform; (*E*) normal and bifurcated navicular.

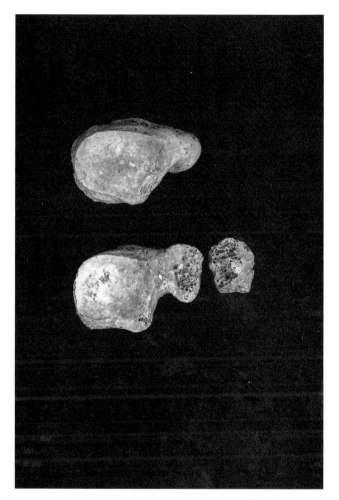

FIGURE F-5.7.2. Tarsal os navicular: left (bottom) adult male, Frankish Corinth, Greece, compared with normal left navicular (top).

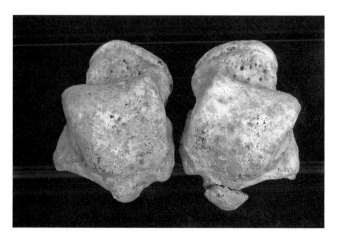

FIGURE F-5.7.3. Tarsal os trigonum: right with normal left talus, adult male, Byzantine Corinth, Greece.

FIGURE F-5.7.4. Tarsal calcaneus secundarius: bilateral, adult female (185), NMNH Terry collection.

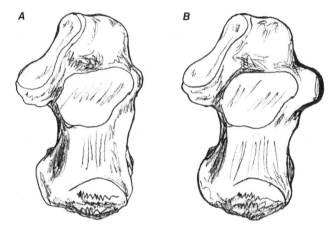

FIGURE F-5.8. Calcaneus hyperplasia peroneal tubercle: (right) (*A*) normal peroneal tubercle; (*B*) enlarged peroneal tubercle (drawn from Dwight 1907).

and plantar segments with the plantar portion the larger. The two segments together are usually larger than a single bone (Burnett and Case 2011; Kjellstrom 2004). The bifurcated navicular follows an oblique separation with the smaller portion on the dorsal–lateral aspect that may override the borders of the second and third cuneiforms (Wiley and Brown 1981).

F-5.8. Tarsal Hyperplasia/ Hypoplasia/Aplasia

This usually impacts only marginal elements, similar to carpals in the hand, but often goes unnoticed in the foot because there is so much variability in tarsal development; it is more apt to be noticed if extreme with segmental disorders.

However, a very rare and unusual isolated hyperplasia has been identified in the form of hyperplasia peroneal tubercle (Fig. F-5.8) on the calcaneus, occurring bilaterally in an adult male (Dwight 1907).

F-6. DIGITAL DEVELOPMENT

Metatarsals and phalanges develop from digital rays in a proximal–distal sequence under the direction of the AER. Primary shaft ossification begins with the middle digital rays, the second, third, and fourth metatarsals, during the ninth embryonic week, followed by primary shaft ossification in the first and fifth metatarsals in the tenth week. Primary ossification in the phalanges begins with the first distal phalanx around the ninth to tenth week, followed by the other distal phalanges. The distal ends of the terminal phalanges ossify directly from the membranous tissue. The proximal phalanges begin to ossify between the eleventh and fifteenth week, followed by the intermediate phalanges. Secondary ossification of the first metatarsal base epiphyses begins in the third year, followed by ossifications of the epiphyseal heads of the other metatarsals between 3 and 5 years, uniting with the shafts between 17 and 20 years. Occasionally, the first metatarsal also forms a separate head epiphysis, and the fifth metatarsal may form a separate ossification for the base tubercle. Epiphyseal ossifications for the bases of the phalanges appear between 2 and 8 years, uniting with the shafts by 18 years.

Sesamoids and extra small ovoid ossicles develop during morphogenesis from mesenchymal tissue condensations separated from the mesenchymal anlage bones of the foot. Usually, most of the extra mesenchymal condensations are absorbed into the developing bone templates, while those developing within tendons at metatarsal–phalangeal junctions evolve into sesamoids, particularly a pair at the plantar margin between the first metatarsal and phalange. Surviving extra ossicles are difficult to identify in paleopathology, except when they unite with major bones.

DIGITAL ANOMALIES

F-6.1. Os Metatarsium and Os Vesalianum

These form from mesenchymal condensations that develop separately from mesenchymal metatarsal anlages during morphogenesis. Os metatarsium forms at the border between the first and second metatarsals and first cuneiform, presenting in cartilage at birth and ossifying around 3 years of age, about the same time as the first cuneiform. The ossicle may remain as a separate entity situated between the three bones and may articulate with one or all three. Usually, it unties with one of the adjacent bones while in contact with one of the other bones (Fig. F-6.1.1C–E). Most often, it unites with the first cuneiform (Fig. F-6.1.2) but can unite with either metatarsal (Fig. F-6.1.3) (Dwight 1907).

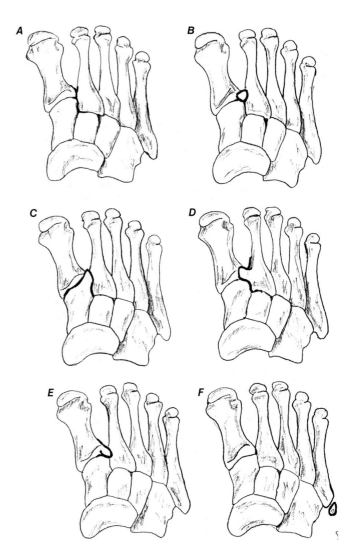

FIGURE F-6.1.1. Os intermetatarsium and os vesalianum: (right) (*A*) normal; (*B*) separate os intermetatarsium; (*C*) os intermetatarsium united with the first cuneiform; (*D*) os intermetatarsium united with the second metatarsal; (*E*) os intermetatarsium united with the first metatarsal; (*F*) os vesalianum at the base of the fifth metatarsal.

Os vesalianum develops as a separate ossification of the fifth metatarsal base tubercle that may unite with the base completely or partially, or remain a separate ossicle (Fig. F-6.1.1F). The separated ossicle attaches to the base by fibrocartilage. This small ossicle can easily be lost but leaves signs of its attachment with pitting at the attachment site (Dwight 1907).

F-6.2. Brachydactyly

This is hypoplasia/aplasia of one or more segments within a digital ray, shortening the affected toe (Fig. F-6.2.1). This is much less common than brachydactyly

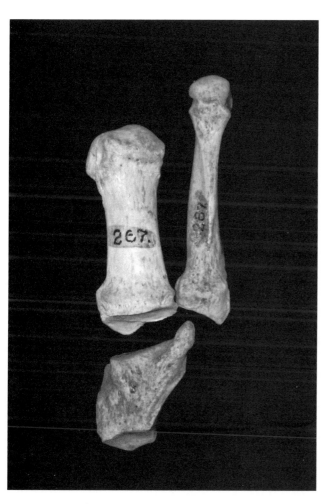

FIGURE F-6.1.2. Os intermetatarsium: unilateral right, united with the first cuneiform, male (267), NMNH Terry collection.

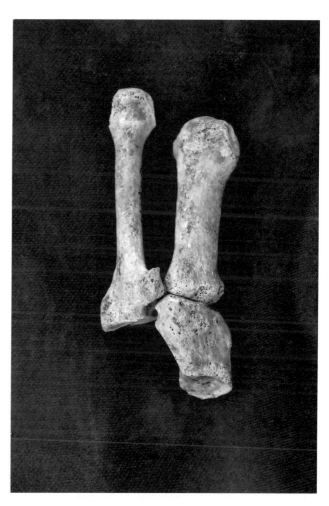

FIGURE F-6.1.3. Os intermetatarsium: unilateral left, united with the second metatarsal, adult female (NMNH 327071), Pueblo Bonito, NM.

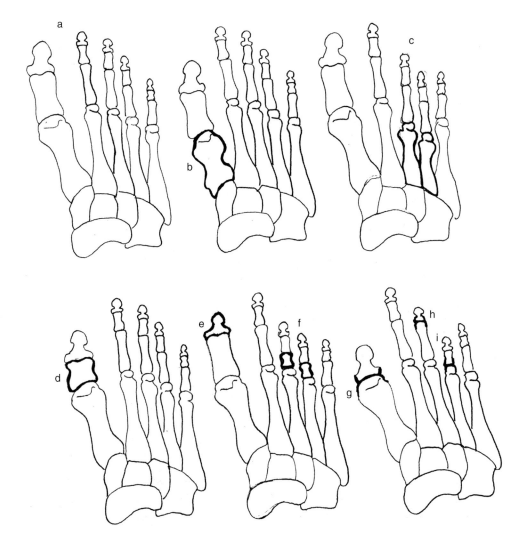

FIGURE F-6.2.1. Brachydactyly types: (right) (a) normal; (b) first metatarsal hypoplasia; (c) third and fourth metatarsal hypoplasia; (d) first proximal phalanx hypoplasia; (e) first distal phalanx hypoplasia; (f) third and fourth middle phalange hypoplasia; (g) aplasia first proximal phalanx; (h) aplasia third middle phalanx; (i) aplasia fourth proximal phalanx.

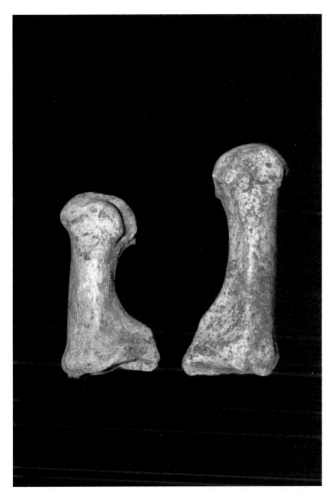

FIGURE F-6.2.2. Brachydactyly: unilateral right first metatarsal with normal left, adult female (NMNH 383150), Mobridge, SD.

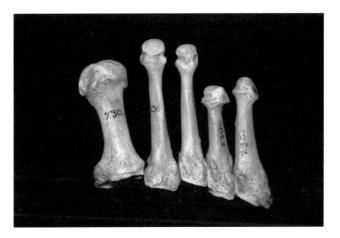

FIGURE F-6.2.3. Brachydactyly: right fourth metatarsal (bilateral), adult male (230), NMNH Terry collection.

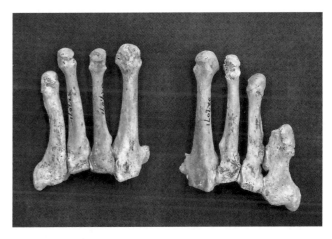

FIGURE F-6.2.4. Brachydactyly: unilateral right fifth metatarsal articulating with MC4 (also bilateral os intermetatarsium), adult female (NMNH 327071), Pueblo Bonito, NM.

affecting finger rays, and much more difficult to decipher when phalanges, especially distal ones, are affected in ancient human remains, not knowing if they were lost during recovery. Also, the variability of size in toe phalanges is considerably more than in the fingers. Most likely forms of brachydactyly to be noted in the foot will be expressed as metatarsal hypoplasia (Figs. F-5.3.4A and F-6.2.2–F-6.2.4), proximal phalangeal hypoplasia, and sometimes hypoplasia of the first digit distal phalanx.

F-6.3. Syndactyly Complex

All or part of adjacent digits fails to separate during morphogenesis. Most often, they are conjoined by soft tissue, known as simple syndactyly. Complex syndactyly involves bone tissue, easily identified in human skeletal remains, varying from conjoined metatarsals to union of distal phalanges (Fig. F-6.3). The most affected digits are the second and third digital rays. Syndactyly

is often associated with other digital anomalies (Kelikian 1974; Temtamy and McKusick 1978).

F-6.4. Symphalangism

This is the absence of interphalangeal joint with failure of segmentation of adjacent phalanges within a single digit (Fig. F-6.4.1). More than one digit in the same foot can be affected, beginning with the lateral side of the foot and progressing to the second digit in order. Most pedal symphalangism occurs in the distal interphalangeal joint (Case and Heilman 2005), unlike the hand that is affected more often by proximal interphalangeal joint symphalangism. Rarely, both interphalangeal joints are involved. The first digit is usually not affected. Symphalangism in the distal interproximal joint of the fifth toe is very common (Fig. F-6.4.2), sometimes with involvement of the fourth toe.

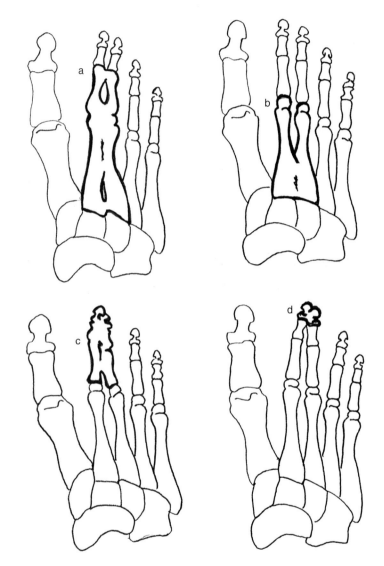

F I G U R E F-6.3. Syndactyly complex: (second and third digital rays) (*A*) conjoined metatarsals–proximal phalanges; (*B*) partially conjoined metatarsals; (*C*) conjoined phalanges; (*D*) conjoined distal phalanges only.

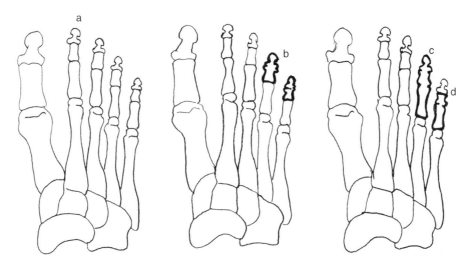

FIGURE F-6.4.1. Symphalangism: (right) (a) normal; (b) united fourth and fifth distal interphalangeal joints; (c) united fourth proximal and distal interphalangeal joints; (d) united fifth proximal interphalangeal joint.

FIGURE F-6.4.2. Symphalangism fifth toe: unilateral right distal interphalangeal joint with normal left, adult male, Frankish Corinth, Greece.

F-6.5. Ectrodactyly

This is agenesis of all or part of one or more digital rays of the forefoot with or without the involvement of the middle foot, caused by disturbances in signals from the AER during morphogenesis for the digital rays. Split (cleft) foot is the most often recognized form of ectrodactyly, often referred to as lobster claw foot that may or may not occur with similar defect of the hand. This phenomenon is frequently bilateral and familial. The middle digital ray is usually absent. There are many different expressions of missing or partially missing parts of digital rays (Figs. F-6.5.1–F-6.5.3). Quite often, remaining digital rays are conjoined (symphalangism), and sometimes can be accompanied by tarsal coalition, and or brachydactyly. The third cuneiform can be missing with the third digital ray. There are many different configurations of split foot that can involve all three middle digital rays, leaving the lateral and medial digits intact, or leaving just a single first digit in place (Blauth and Borisch 1990). Other expressions of ectrodactyly include agenesis of just a single digit, leaving the foot with only four toes (Fig. F-6.5.3D).

F-6.6. Polydactyly

This involves supernumerary or duplicate digital rays. Complete duplication produces an extra, segmented digit, while partial expressions develop from incomplete separation of a digital segment with duplication of segments distal to the bifurcation. Sometimes, the entire

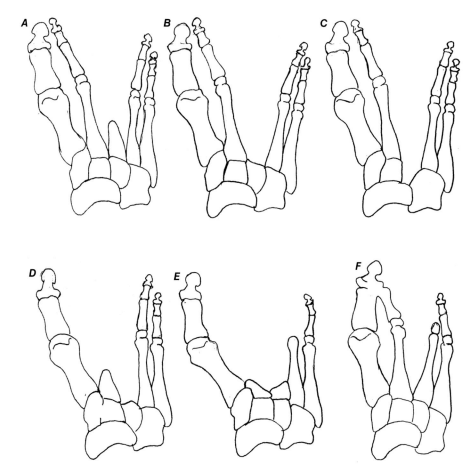

FIGURE F-6.5.1. Ectrodactyly (split foot): (*A*) hypoplasia MT3, phalange agenesis; (*B*) complete agenesis third digital ray; (*C*) complete agenesis third digital ray and third cuneiform; (*D*) incomplete agenesis second digital ray, complete agenesis third digital ray; (*E*) incomplete agenesis MT2 and MT3 with phalange agenesis and fourth digit phalange agenesis; (*F*) symphalangism first and second distal phalanges, incomplete agenesis third digital ray phalanges, complete agenesis fourth digital ray.

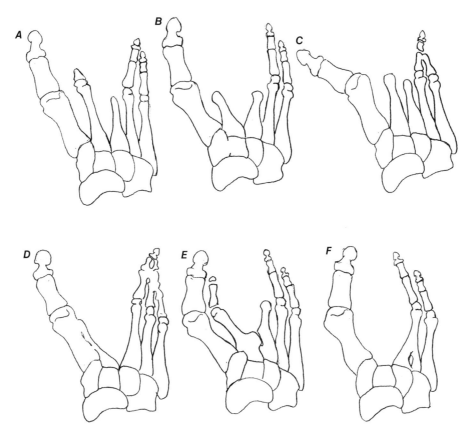

FIGURE F-6.5.2. Ectrodactyly (split foot): (*A*) incomplete agenesis second digit phalanges, hypoplasia MT4 with phalange agenesis; (*B*) hypoplasia MT2 and MT3 with phalange agenesis; (*C*) phalange agenesis second and third digits, hypoplasia MT3, phalangeal symphalangism fourth and fifth digits; (*D*) second digital ray absorbed by MT1, third, fourth, and fifth digit phalangeal symphalangism; (*E*) incomplete union MT2 and MT3 with a wide cleft, partial agenesis second digit distal phalanges, agenesis third digit phalanges; (*F*) complete agenesis second digital ray, MT3 merged with MT4.

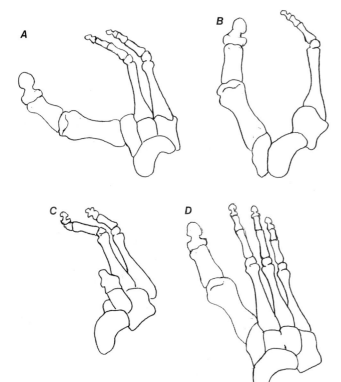

FIGURE F-6.5.3. Ectrodactyly (split foot): (*A*) complete agenesis second and third digital rays, third cuneiform; (*B*) complete agenesis second, third, fourth digital rays, second and third cuneiforms; (*C*) complete agenesis first and second digital rays, first cuneiform, incomplete agenesis second cuneiform, hypoplasia MT3 with phalange agenesis; (*D*) complete agenesis fourth digital ray.

incomplete duplicating segment, especially the metatarsal, appears broader than usual (block segment) or just the distal end is broadened with duplicated following segments. Bifurcated shaft expressions vary from Y shape to branching off at an angle. Rudimentary forms of digits can also develop, usually in soft tissue only. Extra digits generally follow the anterior (preaxial)–

posterior (postaxial) axes during morphogenesis. Very rarely is the central third digital ray affected.

Preaxial polydactyly involves the first digital ray and, on rare occasions, the second digital ray (Fig. F-6.6.1).

Postaxial polydactyly affects the fifth digital ray and, on occasion, the fourth digital ray (Fig. F-6.6.2).

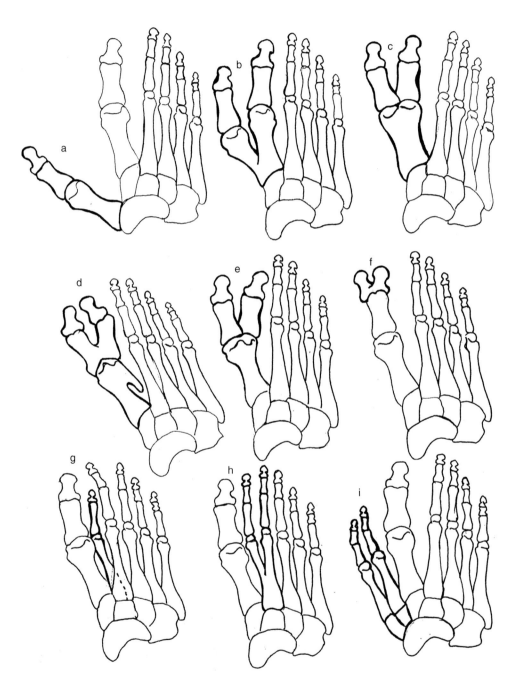

F I G U R E F -6.6.1. Polydactyly—preaxial: (*A*) complete duplication first digit articulating with the navicular; incomplete duplications; (*B*) bifid first metatarsal; (*C*) block first metatarsal, duplicate phalanges; (*D*) incomplete bifurcation first metatarsal and proximal phalanx; (*E*) duplicate first phalanges; (*F*) bifurcated first distal phalanx; (*G*) remnant duplicate second digit; (*H*) bifurcated second metatarsal with duplicate phalanges; (*I*) mirror polydactyly of the second and third digits with remnant cuneiforms.

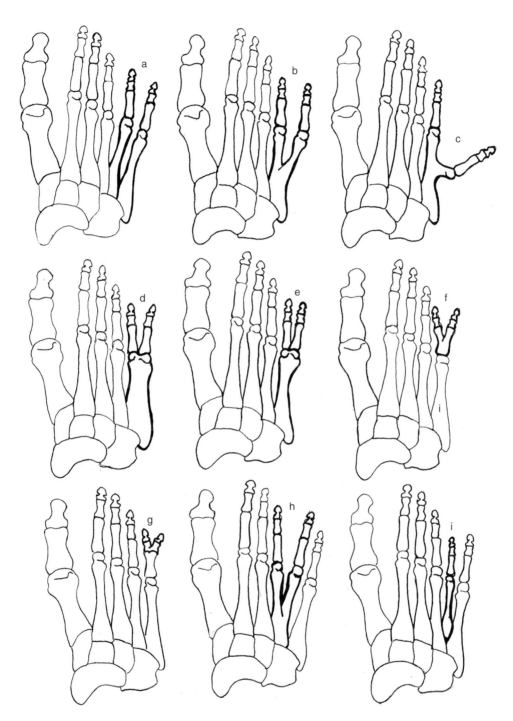

FIGURE F-6.6.2. Polydactyly—postaxial: (*A*) complete duplication fifth digit; incomplete duplications; (*B*) bifurcated fifth metatarsal; (*C*) branching fifth metatarsal; (*D*) block fifth metatarsal with duplicate phalanges; (*E*) duplicate fifth phalanges; (*F*) bifid fifth proximal phalange; (*G*) bifid fifth middle phalange; (*H*) bifurcated fourth metatarsal with duplicate phalanges; (*I*) remnant duplicate fourth digit.

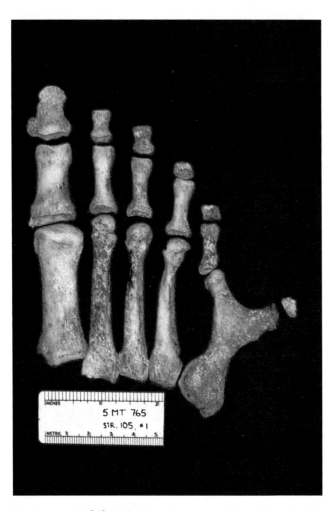

FIGURE F-6.6.3. Polydactyly—postaxial: unilateral right branching fifth metatarsal, adult male, Sand Canyon Pueblo, SW CO (from Bruce Bradley).

The most common form of pedal polydactyly is duplication of the fifth digital ray (Fig. F-6.6.3).

Very rarely, the same foot can have mixed polydactyly with both pre- and postaxial expressions (Barnes 1994b; Case et al. 2006; Watanabe et al. 1992). Extremely rare is a form of mirror foot—mirror polydactyly—one or two extra digital rays along side the medial (Fig. F-6.6.1I) or lateral digits that duplicate digital rays on the other side of the outer digit (Theodorou et al. 2003; Verghese et al. 2007).

LITERATURE CITED

Aalami-Harandi, B. & A. Zahir. 1976. Congenital bifid femur. *Acta Orthopaedica Scandinavica* 47: 419–422.

Anderson, T. 1995. An anomalous medieval parietal bone. *Journal of Paleopathology* 7(3): 223–226.

Anderson, T. 2003. A medieval example of a sagittal cleft or "butterfly" vertebra. *International Journal of Osteoarchaeology* 13: 352–357.

Anderson, T. & A.R. Carter. 1995. The first archaeological case of Madelung's deformity? *International Journal of Osteoarchaeology* 5: 168–173.

Ashley, G.T. 1954. The morphological and pathological significance of synostosis at the manubrio-sternal joint. *Thorax* 9: 159–166.

Ashley, G.T. 1956. The relationship between the pattern of ossification and the definitive shape of the mesosternum in man. *Journal of Anatomy* 90: 87–105.

Aufderheide, A.C. & C. Rodriguez-Martin. 1998. *The Cambridge Encyclopedia of Human Paleopathology*, Cambridge, UK: Cambridge University Press.

Bailey, R.W. 1974. *The Cervical Spine*, Philadelphia: Lea & Febiger.

Barnard, L.B. & S.M. McCoy. 1946. The supracondyloid process of the humerus. *The Journal of Bone & Joint Surgery* 28: 845–850.

Barnes, E. 1994a. *Defects Developmental of the Axial Skeleton in Paleopaleopathology*, Niwot: University of Colorado Press.

Barnes, E. 1994b. Polydactyly in the Southwest. *The Kiva* 59(4): 419–431.

Barnes, E. 2008. Congenital anomalies. In: *Advances in Human Paleopathology*, edited by R. Pinhasi & S. Mays, pp. 329–362. West Sussex: John Wiley & Sons.

Barnes, E. 2012. Developmental disorders in the skeleton. In: *A Companion to Paleopathology*, edited by A.L. Grauer, pp. 380–400. West Sussex, UK: Wiley-Blackwell.

Bell, J. 1951. On brachydactyly and symphalangism. In: *The Treasury of Human Inheritance*, edited by L.S. Penrose, pp. 1–17. Cambridge, UK: Cambridge University Press.

Blackwood, H.J. 1957. The double-headed mandibular condyle. *American Journal of Physical Anthropology* 15: 1–8.

Blauth, W. & N.C. Borisch. 1990. Cleft feet: proposals for a new classification based on roentgenographic morphology. *Clinical Orthopaedics* 258: 41–48.

Boyd, G.I. 1933. Bipartite carpal navicular bone. *The Journal of Bone & Joint Surgery* 20: 455–458.

Brailsford, J.F. 1948. *The Radiology of the Bones and Joints*, Baltimore: Williams & Wilkens.

Brothwell, D.R. 1967. Major congenital anomalies of the skeleton: evidence from earlier populations. In: *Diseases in Antiquity*, edited by D. Brothwell & A.T. Sandison, pp. 423–443. Springfield: Charles C. Thomas.

Brothwell, D.R. 1981. *Digging Up Bones*, Ithaca: Cornell University Press.

Burdi, A.R., T.J. Lawton, & J. Grosslight. 1988. Prenatal pattern emergence in early human facial development. *The Cleft Palate Journal* 25(1): 8–15.

Burnett, S.E. 2011. Hamate-pisiform coalition: morphology, clinical significance, and a simplified classification scheme for carpal coalition. *Clinical Anatomy (New York, N.Y.)* 24(2): 188–196.

Burnett, S.E. & D.T. Case. 2011. Bipartite medial cuneiform: new frequencies from skeletal collections and a meta-analysis of previous cases. *HOMO—Journal of Comparative Human Anatomy* 62(2): 109–125.

Camarda, A.J., C. Deschamps, & D. Forest. 1989. Stylohyoid chain ossification: a discussion of etiology. *Oral Surgery, Oral Medicine, and Oral Pathology* 67(5): 508–514.

Canci, A., E. Marini, G. Mulliri, E. Usai, L. Vacca, G. Floris, & S.M. Borgogononi Tarli. 2002. A case of Madelung's deformity in a skeleton from Nuragic Sardinia. *International Journal of Osteoarchaeology* 12: 173–177.

Carson, W., W.W. Lovell, & T.E. Whitesides. 1981. Congenital elevation of the scapula. *The Journal of Bone and Joint Surgery [A]* 63: 1199–11207.

Case, D.T. 1996. *Developmental Defects of the Hands and Feet in Paleopathology*. MA thesis. Tempe: Arizona State University.

Case, D.T. & J. Heilman. 2005. Pedal symphalangism in modern American and Japanese skeletons. *HOMO—Journal of Comparative Human Biology* 55: 251–262.

Case, D.T., R.J. Hill, C.F. Merbs, & M. Fong. 2006. Polydactyly in the prehistoric American Southwest. *International Journal of Osteoarchaeology* 16: 221–235.

Castriota-Scanderbeg, A. & B. Dallapiccola. 2005. *Advanced Skeletal Phenotypes: From Simple Signs to Complex Diagnoses*, Berlin: Springer.

Cigtay, O.S. & V.J. Mascatello. 1979. Scapular defects: a normal variation. *AJR. American Journal of Roentgenology* 132: 239–241.

Cooper, P.D., J.D. Stewart, & M.S. McCormick. 1988. Development and morphology of the sternal foramen. *The American Journal of Forensic Medicine and Pathology* 9(4): 342–347.

Currarino, G., E. Sheffield, & D. Twickler. 1998. Congenital glenoid dysplasia. *Pediatric Radiology* 28: 30–37.

David, T.J. & R.L. Burwood. 1972. The nature of inheritance of Kirner's deformity. *Journal of Medical Genetics* 9: 430–433.

Dawson, E.G. & L. Smith. 1979. Atlanto-axial subluxation in children due to vertebral anomalies. *The Journal of Bone and Joint Surgery [A]* 61: 582–587.

DeMyer, W. 1967. The median cleft face syndrome. *Neurology* 17: 961–971.

Denninger, H.S. 1931. The pathology of the prehistoric American Indians of the Illinois River Valley. *Illinois State Academy of Science* 24(2): 371–375.

Atlas of Developmental Field Anomalies of the Human Skeleton: A Paleopathology Perspective, First Edition. Ethne Barnes.
© 2012 Wiley-Blackwell. Published 2012 by John Wiley & Sons, Inc.

Dickel, D.N. & G.H. Doran. 1989. Severe neural tube defect syndrome from the early archaic of Florida. *American Journal of Physical Anthropology* 80: 325–334.

Dudor, J.C. 2010. Qualitative and quantitative diagnosis of lethal cranial neural tube defects from the fetal and neonatal human skeleton, with a case study involving taphonomically altered remains. *Journal of Forensic Sciences* 55(4): 877–883.

Dunn, P.M. 1976. The anatomy and pathology of congenital dislocation of the hip. *Clinical Orthopaedics and Related Research* 119: 23–27.

Dwight, T. 1907. *A Clinical Atlas: Variations of the Bones of the Hands and Feet*, Philadelphia: J.B. Lippincott Co.

Eijgelaar, A. & J.H. Bijtel. 1970. Congenital cleft sternum. *Thorax* 25: 490–498.

Epstein, B.S. 1976. *The Spine: A Radiological Text and Atlas*. 4th ed. Philadelphia: Lea & Febiger.

Ferguson, A.B. 1968. *Orthopedic Surgery in Infancy and Childhood*. 3rd ed. Baltimore: Williams & Wilkins Co.

Flatt, A.E. & V.E. Wood. 1975. Rigid digits or symphalangism. *The Hand* 7: 197–213.

Fraser, F.C. 1963. Harelip and cleft palate. In: *Birth Defects*, edited by M. Fishbein, pp. 235–245. Philadelphia: J.B. Lippincott.

Freni, S.C. & W.F. Zapisek. 1991. The cleft palate. *The Cleft Palate––Craniofacial Journal* 28(4): 338–345.

Garn, S.M., A.R. Burdi, & W. Babler. 1976. Prenatal origins of carpal fusions. *American Journal of Physical Anthropology* 45: 203–208.

Gillespie, R. & J.P. Torode. 1983. Classification and management of congenital abnormalities of the femur. *The Journal of Bone & Joint Surgery* B65(5): 557–568.

Gladykowska-Rzeczycka, J.J. & T. Mazurek. 2009. A rare case of forearm hypoplasia from 18th-century Gdansk, Poland. *International Journal of Osteoarchaeology* 19: 726–734.

Golthamer, C.R. 1957. Duplication of the clavicle. *Radiology* 68: 576–578.

Goodman, R.M. & R.J. Gorlin. 1983. *The Malformed Infant and Child: An Illustrative Guide*, New York: Oxford University Press.

Gossman, J.R. & J.J. Tarsitano. 1977. The styloid-stylohyoid syndrome. *Journal of Oral Surgery (American Dental Association: 1965)* 35(7): 555–560.

Gragon, D.P., S.M. Love, & J.A. Ogdan. 1987. Congenital malformations of the lower extremities. *The Orthopedic Clinics of North America* 18: 537–554.

Green, J.P. & W. Waugh. 1968. Congenital lateral dislocation of the patella. *The Journal of Bone & Joint Surgery* B50(2): 285–289.

Greenberg, D.P.M. 1981. Congenital vertical talus and congenital calcaneovalgus deformity: a comparison. *The Journal of Foot Surgery* 20: 189–193.

Griffin, P. & F. Rand. 1988. Static deformities. In: *The Foot Volume I*, edited by B. Helal & D. Wilson, pp. 385–410. Edinburgh: Churchill Livingstone.

Gruneberg, H. 1963. *The Pathology of Development: A Study of Inherited Skeletal Disorders in Animals*, New York: John Wiley & Sons.

Gruneberg, H. 1964. The genesis of skeletal anomalies. In: *Congenital Malformations*, edited by M. Fishbein, pp. 219–223. New York: International Medical Congress.

Guttenag, A.R. & J.K. Salwen. 1999. Keep your eyes on the ribs: the spectrum of normal variants and diseases that involve the ribs. *Radiographics* 19: 1125–1142.

Harris, R.I. 1959. Congenital anomalies. In: *Modern Trends in Diseases of the Vertebral Column*, edited by R. Nassim & H.J. Burrows, pp. 29–66. New York: Paul Hoeber.

Hauser, V. & G.F. De Stefano. 1989. *Epigenetic Variants of the Human Skull*, Stuttgart: Schweizerbart.

Hoffman, J.M. 1976. Enlarged parietal foramina—their morphological variation and use in assessing prehistoric biological relationships. In: *Studies in California Paleopathology*, edited by M.J. Hoffman & L. Brunker, pp. 41–64. Berkeley. University of California, Department of Anthropology.

Honeij, J.A. 1920. Cervical ribs. *Surgery, Gynecology and Obstetrics* #0: 481–493.

Hrdlicka, A. 1933. Seven prehistoric American skulls with complete absence of external auditory meatus. *American Journal of Physical Anthropology* 17(3): 355–379.

Hutchinson, D.L., C. Denise, H.J. Daniel, & G.W. Kalmus. 1997. A reevaluation of the cold water etiology of external auditory exostoses. *American Journal of Physical Anthropology* 103: 417–422.

Jerome, J.T.J., M. Varghese, & B. Sankaran. 2009. Congenital patellar syndrome. *Romanian Journal of Morphology and Embryology* 50(2): 291–293.

Jones, G.B. 1964. Delta phalanx. *Journal of Bone and Joint Surgery* B46: 226–228.

Kaban, L.B., J.B. Mulliken, & J.F. Murray. 1981. Three-dimensional approach to analysis and treatment of hemifacial microsomia. *The Cleft Palate Journal* 18(2): 90–99.

Kawashima, T. & H.K. Uhtoff. 1990. Prenatal development around the sustentaculum tali and its relation to talocalcaneal coalitions. *Journal of Pediatric Orthopaedics* 10: 238–243.

Keim, H.A. & R.N. Hensinger. 1989. Spinal deformities. *Clinical Symposia (Summit, N.J.: 1957)* 41(4): 10–11; 16–17.

Kelikian, H. 1974. *Congenital Deformities of the Hand and Forearm*, Philadelphia: W.B. Saunders.

Kim, S.J. & B.H. Min. 1994. Congenital absence of the acromion. A case report. *Clinical Orthopaedics and Related Research* 300: 117–119.

Kim, S.J., J.S. Kim, H.J. Kim, & H.W. Yu. 1998. Bilateral unfused coracoid process: a report of a case. *Journal of Korean Medical Science* 13(5): 563–565.

Kjellstrom, A. 2004. A case study of os cuneiforme mediale bipartum from Sigtuna, Sweden. *International Journal of Osteoarchaeology* 14: 475–480.

Larsen, W.J. 2001. *Human Embryology*, Philadelphia: Churchill Livingstone.

Le Minor, J.-M. & O. Trost. 2004. Bony ponticles of the atlas (C1) over the groove for the vertebral artery in humans and primates: polymorphism and evolutionary trends. *American Journal of Physical Anthropology* 125: 16–29.

Lemire, R.J. 1988. Neural tube defects. *Journal of the American Medical Association* 259(4): 558–562.

Levine, M.A. 1950. Patella cubiti. *The Journal of Bone & Joint Surgery* 32A(3): 686–687.

Little, J.W. & J. Jakobsen. 1973. Origin of the globulomaxillary cyst. *Journal of Oral Surgery (American Dental Association: 1965)* 31: 188–195.

Mac-Tniong, J.M., S. Leduc, & H. Labelle. 2005. Complete bilateral agenesis of the ilium in a seven-year-old ambulatory girl: a case report. *Spine* 30(4): 420–423.

Manashil, G. & S. Laufer. 1979. Congenital pseudoarthrosis of the clavicle. *AJR. American Journal of Roentgenology* 132: 678–679.

Mann, G.E. 1984. *The Torus Auditivus: A Re-appraisal and Its Application to a Series of Ancient Egyptian Skulls*. M.Phi. dissertation. Cambridge, UK: Cambridge University.

McCormick, S.U., S.A. McCormick, R.W. Graves, & R.G. Pifer. 1989. Bilateral bifid mandibular condyles. *Oral Surgery, Oral Medicine, & Oral Pathology* 68(5): 555–557.

Merbs, C.F. 2004. Sagittal clefting of the body and other vertebral developmental errors in Canadian Inuit skeletons. *American Journal of Physical Anthropology* 123: 236–249.

Mladick, R.A., C.E. Horton, J.E. Adamson, & J.M. Carraway. 1974. Medial, lateral, and transverse clefts. In: *Symposium on Management of Cleft Lip and Palate and Associated Deformities*, Volume 8, edited by N.G. Georgiade. St. Louis: C.V. Mosby.

Moore, K.L. 1985. *Clinical Oriented Anatomy*, Baltimore: Williams & Wilkins.

Myrianthopoulos, N.C. & M. Melnick. 1989. Studies in neural tube defects. I. Epidemiologic and etiologic studies. *American Journal of Medical Genetics* 26(4): 783–796.

Oner, F.C. & H.R. de Vries. 1994. Isolated capitolunate coalition. *The Journal of Bone & Joint Surgery* 76B: 845–846.

Ortner, D.J. 2003. *Identification of Pathological Conditions in Human Skeletal Remains*, Amsterdam: Academic Press.

Ostrum, R.F., R.R. Betz, M. Clancy, & H.H. Steel. 1987. Bifurcated femur with a normal tibia and fibula. *Journal of Pediatric Orthopaedics* 7(2): 224–226.

Ponseti, I.V. 1978. Morphology of the acetabulum in congenital dislocation of the hip. *The Journal of Bone & Joint Surgery* 60A: 586–599.

Regan, M., D.T. Case, & J. Cleaves-Brundige. 1999. Articular surface defects in the third metatarsal and third cuneiform: non-osseous tarsal coalition. *American Journal of Physical Anthropology* 109(1): 53–66.

Resnik, C.S., J.D. Grizzard, B.P. Simmons, & I. Yaghmai. 1986. Incomplete carpal coalition. *AJR. American Journal of Roentgenology* 147: 301–304.

Rubin, G., R. Scienza, A. Pasqualin, L. Rosta, & R. Da Pian. 1989. Craniocerebral epidermoids and dermoids: a review of 44 cases. *Acta Neurochirurgica* 97: 1–16.

Ruge, D. & L.L. Wiltse. 1977. *Spinal Disorders: Diagnosis and Treatment*, Philadelphia: Lea & Febiger.

Sachs, J. & G. Degenshein. 1948. Patella cubiti. *Archives of Surgery* 57(5): 675–680.

Sadler, T.W. 2006. *Langman's Medical Embryology*, Philadelphia: Lippincott Williams & Wilkins.

Schafner, W.G., M.K. Hine, & B.M. Levy. 1983. *A Textbook of Oral Pathology*. 4th ed. Philadelphia: W.B. Saunders.

Scheie, H.G. & D.M. Albert. 1977. *Textbook of Opthalmology*. 9th ed. Philadelphia: W.B. Saunders.

Schmorl, G. & H. Junghanns. 1971. *The Human Spine in Health and Disease*. 2nd ed. New York: Grune & Stratton.

Schultz, R.E. & F.C. Theisen. 1989. Bilateral coronoid hyperplasia. *Oral Surgery, Oral Medicine, & Oral Pathology* 68(1): 23–26.

Selby, S., S.M. Garn, & V. Kanareff. 1955. The incidence and familial nature of a bony bridge on the first cervical vertebra. *American Journal of Physical Anthropology* 13: 129–142.

Settle, G.W. 1963. The anatomy of congenital talipes equinovarus. *The Journal of Bone & Joint Surgery* A45: 1341–1353.

Shapiro, R. 1972. Anomalous parietal sutures and the bipartite parietal bone. *The American Journal of Roentgenology, Radium Therapy, and Nuclear Medicine* 115(3): 567–577.

Shapiro, R. & F. Robinson. 1976. Anomalies of the craniovertebral border. *AJR. American Journal of Roentgenology* 127: 281–287.

Silva, A.M. 2005. Non-osseous calcaneonavicular coalition in the Portuguese prehistoric population: report of two cases. *International Journal of Osteoarchaeology* 15: 449–453.

Stafne, E.C. 1942. Bone cavities situated near the angle of the mandible. *The Journal of the American Dental Association* 29: 1969–1972.

Stark, P., G.E. Watkins, H.E. Hildebrant-Stark, & R.D. Dunbar. 1987. Episternal ossicles. *Radiology* 165(1): 143–144.

Swanson, A.B., G. de Goot Swanson, & K. Tada. 1983. A classification for congenital limb malformation. *The Journal of Hand Surgery* 8: 693–702.

Temtamy, S.A. & V.A. McKusick. 1978. *The Genetics of Hand Malformation*, New York: Alan R. Liss.

Theodorou, S.J., D.J. Theodorou, & S. Farooki. 2003. An unusual case of polydactyly of the foot. *AJR. American Journal of Roentgenology* 181(6): 1721–1722.

Tsou, P.M., A. Yau, & A.R. Hodgson. 1980. Embryogenesis and prenatal development of congenital vertebral anomalies and their classification. *Clinical Orthopaedics* 152: 211–231.

Tuncay, L., R. Akpinar, O. Unal, & A. Aydinlioglu. 2001. Congenital complete entire carpal fusion with massive carpometacarpal coalition. *Eastern Journal of Medicine* 6(2): 53–54.

Twigg, H.L. & R.C. Rosenbaum. 1981. Duplication of the clavicle. *Skeletal Radiology* 6: 281.

Urunuela, G. & R. Alvarez. 1994. A report of Klippel-Feil syndrome in prehistoric remains from Cholula, Puebla, Mexico. *Journal of Paleopathology* 16(2): 63–67.

Usher, B.M. & M.N. Christensen. 2000. A sequential developmental field defect of the vertebrae, ribs and sternum, in a young woman of the 12th century A.D. *American Journal of Physical Anthropology* 111(3): 355–367.

Verghese, B., H. Shah, G. Rebello, & B. Joseph. 2007. Pre-axial mirror polydactyly associated with tibial deficiency: a study of the patterns of skeletal anomalies of the foot and leg. *Journal of Children's Orthopaedics* 1(1): 49–54.

Viladot, A. 1988. Local congenital disorders. In: *The Foot Vol. I*, edited by B. Helal & D. Wilson, pp. 235–264. Edinburgh: Churchill Livingstone.

Warkany, J. 1971. *Congenital Malformations*, Chicago: Yearbook Medical Publications, Inc.

Watanabe, H., S. Fujita, & O. Ichiro. 1992. Polydactyly of the foot: an analysis of 265 cases and a morphological classification. *Plastic and Reconstructive Surgery* 89: 856–877.

Webb, S.G. & A.G. Thorne. 1985. A congenital meningocoele in prehistoric Australia. *American Journal of Physical Anthropology* 68: 525–533.

Weinberg, M.V. & T.B. Van De Mark. 1972. Midline cleft of the mandible: review of the literature and report of a case. *Journal of Oral Surgery (American Dental Association: 1965)* 30(2): 143–148.

White, J.C., M.H. Poppel, & R. Adams. 1945. Congenital malformations of the first thoracic rib. *Surgery, Gynecology & Obstetrics* 81: 643–659.

Wiley, J.J. & D.E. Brown. 1981. The bipartite tarsal scaphoid. *The Journal of Bone & Joint Surgery* B63(4): 583–586.

Willis, T.A. 1929. An analysis of vertebral anomalies. *American Journal of Surgery* 6: 163–168.

Wilson, C.B. & H.A. Norrell. 1966. Congenital absence of a pedicle in the cervical spine. *AJR. American Journal of Roentgenology* 97(3): 639–647.

Wolf, J., K. Mattila, & O. Ankkuriniemi. 1986. Development of a Stafne mandibular bone cavity: report of a case. *Oral Surgery, Oral Medicine, and Oral Pathology* 61: 519–521.

Wolfgang, G.L. 1984. Complex congenital anomalies of the lower extremities: femoral bifurcation, tibial hemimelia, and diastasis of the ankle. *The Journal of Bone & Joint Surgery* A66(3): 453–458.

Wright, L.E. 2011. Bilateral talipes equinovarus from Tikal, Guatemala. *International Journal of Paleopathology* 1: 55–62.

Yates, J.R., M.A. Ferguson-Smith, A. Shenkin, R. Guszman-Rodriguez, M. White, & B.J. Clark. 1987. Is disordered folate metabolism the basis for the genetic predisposition to neural tube defects? *Clinical Genetics* 31(5): 279–287.

Zguricas, J., P.F. Dijkstra, E.S. Gelsema, P.J. Snijders, H.P. Wusterfeld, H.W. Venema, S.E. Hovius, & D. Lindbout. 1997. Metacarpophalangeal pattern (MCPP) profile analysis in a family with triphalangeal thumb. *Journal of Medical Genetics* 34(1): 55–62.

Zimmerman, A.W. & C.B. Lozzia. 1989. Intersection between selenium and zinc in the pathogenesis of anencephaly and spina bifida. *Zeitschrift für Kinderchirurgie* 44(Suppl. 1): 48–50.

INDEX

Atlas of Developmental Field Anomalies of the Human Skeleton: A Paleopathology Perspective, First Edition. Ethne Barnes.
© 2012 Wiley-Blackwell. Published 2012 by John Wiley & Sons, Inc.